AMERICA'S FASTEST GROWING HOBBY!

This year, twenty *billion* beer cans will roll off the assembly lines of America's manufacturers!

And an estimated fifteen thousand Americans will start snatching them up! And not for the foaming amber contents, either–those fifteen thousand citizens are crazy for the colorful metal containers themselves. They're beer can collectors, a new breed of hobbyist, and their ranks are growing at a staggering rate!

They've got a national organization; they hold yearly can-ventions; they trade by mail, by newsletter, by everything but pony express. They find *valuable antique cans* in abandoned liquor stores, in vacant lots, in the walls of buildings, and on each other's carefully arranged display shelves. You'll meet many of them, and their fascinating collections, in the fun-and-photo-filled pages of this book!

That's right! BEER CAN COLLECTING tells you where, why, and how it's done! We guarantee that after you've read it, you may or may not become a collector–but a can of beer will *never* look the same!

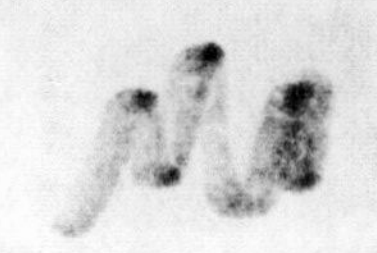

America's fastest growing hobby

by

LEW CADY

tempo
books

GROSSET & DUNLAP
A FILMWAYS COMPANY
Publishers • New York

Dedicated to Lonnie Smith
(With hope that maybe now
he will trade me that Pikes Peak Malt Liquor!)

Tain't easy to put a book like this together without a lot of help and helpers, so I didn't try. I am grateful to *Jim McCoy* for his many hours of research, advice, and criticism, and to *Ron Moermond* for allowing his brain to be picked (and his fabulous collection to be photographed). *Larry Wright* edited the BCCA News Report for many years, and that supplied me with a lot of material. Mile Hi Chapter members *Bob Stutzman, Lonnie Smith,* and *Dick Veit* assisted with the really arcane technical aspects of collecting. *Bob Myers* allowed me to make use of his fine booklet, "Some Thoughts ... on the Beercan's First Thirty-Five Years." *Tom Williams* provided a great deal of information on beer can restoration, and *Phil Chaves* illustrated it. *Bonnie Grauel* made most of the words appear on paper. *Kathy Heider* did the rest. *Don Kibbe* delinted all of the photographs, *Leslie Cady* did everything she could to make sure everything was right, and *Audrey Cady* still wonders where she went wrong.

Cover photo of John Ahren's collection by
Sam Nocella: Philadelphia Bulletin

ISBN: 12535-8 (Tempo Books Edition)
ISBN: 13513-2 (Tempo Library Edition)
Library of Congress Catalog Card Number: 76-13375

Published simultaneously in Canada
Printed in the United States of America

BEER CAN
COLLECTING

Contents

"Many beer can collectors seem to think beer cans are the one bit of madness in their lives. I often feel that beer cans are the sanest part of my existence."

Bill Christensen, Beer Can Collector
Madison, New Jersey

1

Who collects beer cans... and why?

"My mother thinks I'm off my rocker."

Ed Berkler, Beer Can Collector
Alamogordo, New Mexico

What do most men look forward to after a hard day's work?

A nice, cold, can of beer.

But not Ed Berkler. By day, he's a pharmacist. But when he comes home at night, he wants to come home to a nice *empty* can of beer.

Berkler, you see, is one of the 15,000 Americans—men, women and children—whose thirst for empty beer cans is insatiable. So far he has collected over 2,000 of them—all different.

Ed likes nothing better than to receive yet another addition to his collection. Packages of beer cans arrive every few days at the Berkler house—sent by collectors across the nation—collectors with whom Ed trades beer cans.

Beer cans? Aren't they the things that every slob in a T-shirt is supposed to have clutched in his hands, the butt of many a cartoon—and the serious concern of every ecology-minded person from Vancouver to the Florida Keys?

The very same. This year, some 20,000,000,000 of them will roll off the lines of America's can manu-

FROM A TO Z. America's brewers have produced beer in cans for every letter of the alphabet but X.

facturers. And well over a million of them *won't* become litter, *won't* wind up in a landfill, *won't* be recycled. Instead, they'll end up in the homes of America's 15,000 beer can collectors to be displayed proudly for all to see who care to look.

Fact is, more and more people are caring about beer cans every day. And they're starting collections at such a heady rate that nobody really knows how big beer can collecting is—or how big it will become. "It's the country's fastest-growing hobby," says *Business Week*. Who knows? We may look back on it as one of the real phenomena of the decade and it may wind up as commonplace as the collecting of coins and stamps.

As newspaperman Bob Ewegen put it: "When was the

last time you really enjoyed licking a stamp before you stuck it in your collection? Well? Don't you wish you collected beer cans instead?"

And the problem with collecting coins—money you'll never spend—needs no further discussion.

Besides being able to enjoy the contents of a beer can before you set it up on your display shelf (or trade it with another collector for a can you don't have), there are several other things that make beer can collecting a good candidate for The Perfect Hobby.

First of all, there's an unending supply of cans to collect and trade. Not only are the breweries turning them out at an ever-increasing rate, anybody—even if he can't afford the price of a sixpack—can find them by the bushel almost anywhere he goes. As longtime collector Gil Brennell says, "It's a wonderful—if slightly crazy—hobby that all can enjoy whether young or old; rich or poor; male or female; pretty, handsome, or not so pretty or handsome."

If you fit any of those descriptions—and are slightly crazy—you have all the qualifications you need to join in the pursuit of cylindrical metal beer containers.

Just like the bricklayer in St. Louis. And the State Trooper in Cedar Falls. And the geography professor in South Carolina. And the fireman in Chicago. And the Master Sergeant at Tyndall Air Force Base. And the Linotype operator in Western Australia. And the cartographer in Colorado. And the priest in Buffalo, Illinois. And Broderick Crawford. And thousands of students in grade schools, high schools, and colleges all over America.

(College, by the way, is where Steve Baumgartner got his start. He began collecting cans at Purdue. He's now playing defensive end for the New Orleans Saints and calling up fellow collectors in every NFL city he visits, to see what they've got to trade.)

As the bricklayer, the fireman, Steve Baumgartner, and all the rest have discovered, however, there's more to

TODAY, these are the nation's five largest brewers. This is what their first beer cans looked like.

the hobby than collecting beer cans. There are people, too. Although they form as diverse a group as can possibly be imagined, beer can collectors have one thing in common besides their slightly ludicrous hobby. None of them takes himself very seriously.

When Illinois funeral director Shorty Hotz found out there was another mortician/collector up in Rolla, North Dakota, he said, "I'd sure like to get together with him some time and have a cold one!"

Nevertheless, the beer can's appeal as a collectible often has nothing to do with the fact that it once contained beer. In fact, many collectors don't drink beer. (Many aren't old enough.) As Bob Dewitz, a mailman—and longtime collector—in Williamsville, New York, puts it: "They're great little works of art!"

It's true. Call it pop-top art if you like, but the beer can stands head (no pun intended) and shoulders above the Campbell's Soup can as an item of nostalgia and a symbol of our times. To the archeologists of the future, it may well be one of the most important artifacts of civilized man in the Twentieth Century. Of course, they won't have to dig far for them; they'll find rows and rows on shelves in basements throughout the land.

Why beer cans—and not beer bottles? Beer bottles have been around a lot longer than beer cans, so you'd think that *they'd* be the real treasures. Collectors don't see it that way, though, because most bottles carry no imprint of the brewery they came from. They're interchangeable. And many breweries which reuse bottles merely wash off their competitor's label and substitute their own. So anybody who has a supply of old beer labels (and there are lots of them around) can create what looks like an old beer bottle. Besides, it would be easy to have labels for old beers printed up; it's much, *much* harder to counterfeit a beer can.*

---

*If, however, you'd like to see the world's largest collection of beer bottles and labels, don't miss Ernie Oest's museum/saloon, Memories of Beer & Brewing, in Port Jefferson, N.Y. Open Friday, Saturday, Sunday, Monday from 3 p.m. until midnight. No admission charge.

Even Mother Nature seems to want us to collect beer cans, as one big California beer bottle collector found out when he lost a whole bunch of them in an earthquake in 1970. He's no longer a big California beer bottle collector. Now he's a big California beer *can* collector.

In the long run, beer cans may be the rarer collectibles, too. Although a bottle will still be a bottle even if it's left outdoors for thousands of years, a Penn State scientist estimates that, outdoors, an aluminum beer can will break down to dust-size particles of aluminum oxide—and disappear—in just 500 years. A steel can will last only about 100 years.

A good can, though, will disappear a lot faster than that if a collector's eyes fall on it!

2

It all started on January 24, 1935—maybe

> "Most said you would not live long. And through the years you have always been treated kindly. You've been punched, pulled on, kicked, bent double, misprinted and thrown into ditches to fade and rust. Some have even tried to ban you from their states. But you have stood up to the test. And now people are discovering your true beauty, dignity, personality and depth. Be proud, whether you're new, old, rusty, or faded. You're a Beer Can!"
>
> Larry Wright, Beer Can Collector
> St. Louis, Missouri

The history of the beer can is a war story.

It's the story of a mighty battle between those who had put beer in bottles for many years—and those who had the audacity to propose that beer could also be put in cans. The battle raged fiercely in the thirties and forties and has only recently settled down to a semi-peaceful coexistence.

Back in the early 1930s, the three big can companies, American Can, Continental Can, and National Can, offered some strong arguments in favor of providing brew to the public in metal containers. After all, cans stacked better than bottles. They took up much less space. They cooled faster. They weighed less, hence lower shipping charges, and they didn't break. No deposit was needed,

EVEN IN 1945, Blatz cans carried an eight-point argument that was pro-can and anti-bottle.

because cans didn't have to be returned in those days before recycling. No rewashing was necessary; nor did the consumer have to worry about whether a can had last been used to store kerosene or rat poison or some kid's spider collection. And besides all of that, cans completely shielded their contents from the sun, eliminating that disagreeable taste that brewers call "skunkiness."

The bottle boys fought back with their own battery of arguments: Cans, being one-way containers, were more expensive. Cans required a can opener which poked dirt into the beer. Cans may have cooled faster, but they got *warm* faster, too. Cans had no neck, so you were likely to get suds in your eye when you opened one. And—the real crux of the argument—cans were made of metal which is a well-known enemy of beer and is guaranteed to foul the taste of even the finest brew.

The can people had to admit that they were vulnerable on this last point. So they went to work on the development of a substance they could line their cans with to keep the beer from coming into contact with the metal. After nine years, American Can and Union Carbide came up with a plastic coating they called Keglining and the war was on in earnest.

According to *Fortune* (January, 1936), Anheuser-Busch and Pabst had both experimented with canning as early as 1929, but were hesitant to take the commercial plunge.

Not until 1935 could the canners convince a commercial brewery to jump in. It started the year before when American Can engaged the Krueger Brewery of Newark, New Jersey in three months of negotiations. American offered Krueger all kinds of financial guarantees to get them to try canned beer and ale on a test market.

Finally, Krueger came around. Richmond, Virginia was selected as the city for the experiment. And on January 24, 1935, it happened. Krueger Ale appeared for sale in—of all things—a *can*. It was a flattop can*, and

---

*Collectors now call these cans, which require a churchkey, "flats" or "flattops."

KRUEGER STARTED IT ALL back in 1935 (or 1933). This is one of the first Krueger cans to come on the market. Yes, it's rare. Very.

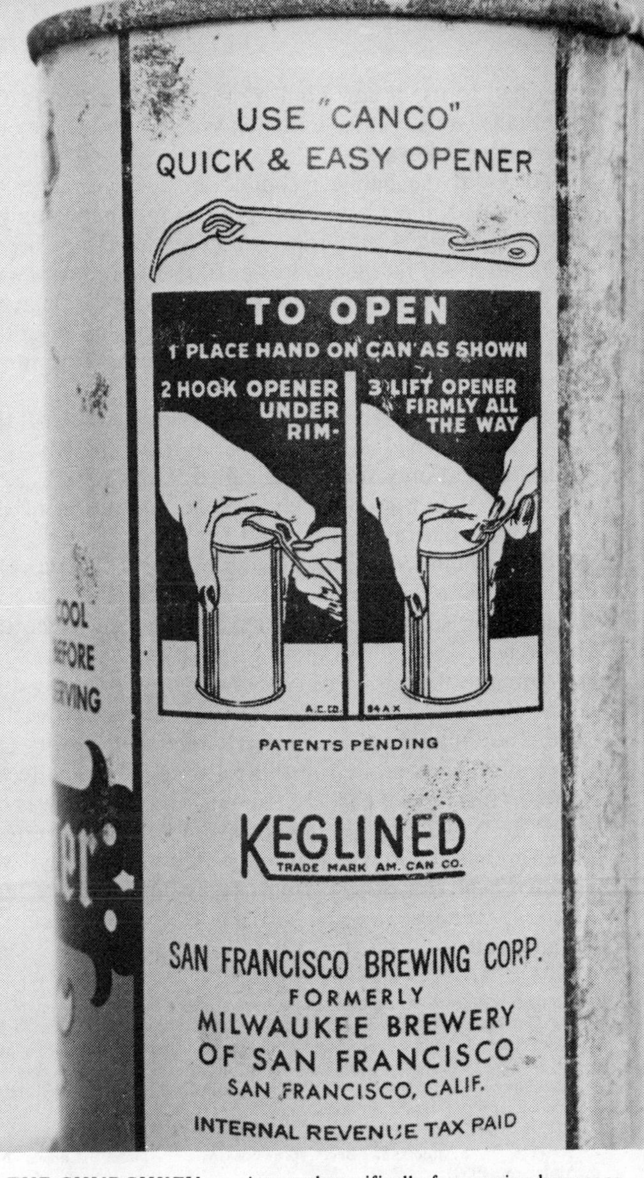

THE CHURCHKEY was invented specifically for opening beer cans. The first beer cans carried step-by-step directions for the benefit of those who didn't know how to use the newfangled contraptions.

you had to use a new kind of opener—invented just for beercans—to get at its contents. We all know these openers as churchkeys.*

How did the public respond? By June, Krueger was running at 550 per cent of its pre-can production and the brewing world went crazy. People *would* buy beer in cans, after all! Suddenly, lots and lots of breweries wanted to get their beer into cans. But American Can played it cool. They turned away the little brewers and saved their canmaking equipment for the big guys—the powerful national brewers.

The first to sign was Pabst, who introduced their canned beer in July. But only, at first, in Rockford, Illinois. And only something called Pabst Export. Some say they didn't want to chance the reputation of Blue Ribbon by putting it into cans.

By September, Schlitz had joined the battle. Only Schlitz came in Continental's can, which had a cone-shaped top so it could be filled and capped by bottling machines.

Among collectors, this type of can is particularly desirable, probably because the cone top shape "proves" its age. In actuality, a few brewers were still putting their beer into conetops well into the 1950's. Many collectors refer to this type of can as a "spout," a term that was first used by Will Anderson in his book, "Beers, Breweries & Breweriana" (1969).† Will says that he and the well-known beer can expert, Bob Myers, invented the term. Today, general usage has made both terms interchangeable.

* Another serious student of beer can collecting, Jim Van Order, maintains that the first beer to be put in cans was actually Krueger Special Beer, a 3.2% product that he says was produced—but just 2,000 cans of it—in December of 1933. Van Order discovered this tidbit of information while reading a newspaper from November of 1933. It is not known, however, whether the beer was actually put on the market.

† Sonja & Will Anderson, "Beers, Breweries & Breweriana" (Carmel, N.Y.: Anderson, 1969).

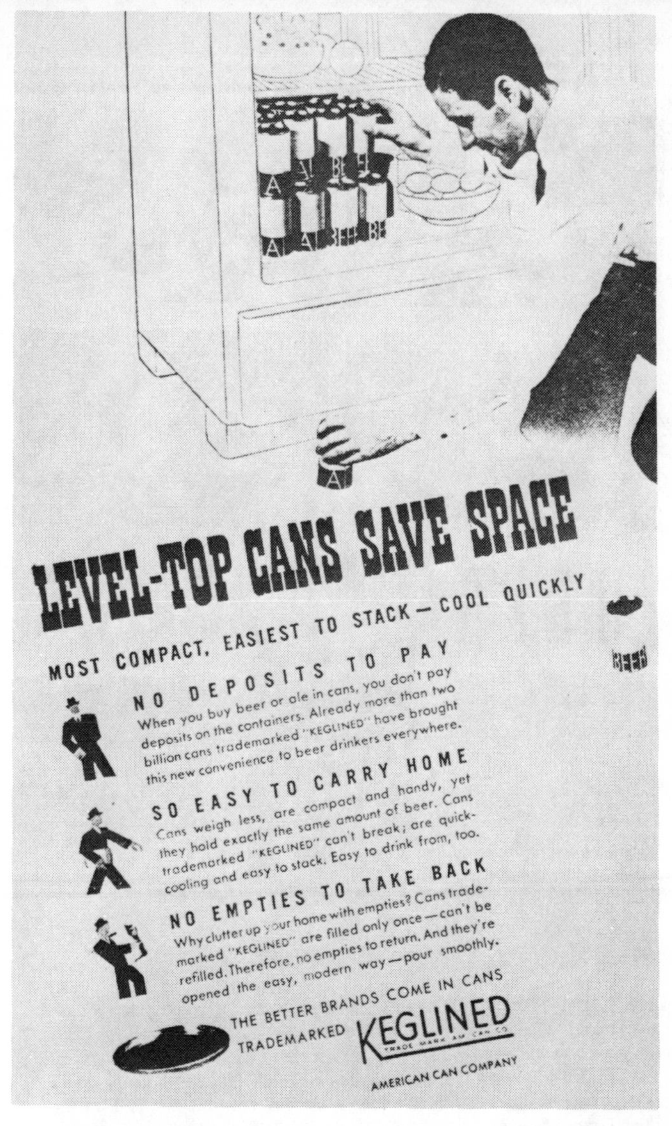

IN 1940, this ad appeared in *Newsweek*. It was just one of American Can Company's many shots in the war between cans and bottles that was raging then.

..it's a picnic

wherever you are, Beer in Cap Sealed Cans

CARTON HOLDS A CASE; HALF THE WEIGHT—HALF THE SPACE

DRINK RIGHT FROM THE CAN; NO EMPTIES TO RETURN

OPENS LIKE A BOTTLE; POURS LIKE A BOTTLE

CLEAN . . . BRINGS BREWERY FLAVOR

BEER *in* CAP SEALED CANS

GOOD BEER goes with good times like crackers go with cheese.

This can brings you the true delicious flavor of draught beer, and is so wonderfully convenient as well. No special opener required; drink right from the clean protected top; a whole carton (full case) is light and takes up little space; no deposits, no bother about returning empties. Remember, for full enjoyment of beer outdoors—or anywhere—insist on beer in Cap Sealed Cans.

EXCLUSIVE FEATURES

Opens like a bottle . . . easy to pour from . . . lined after the can is made . . . cap keeps pouring and drinking surface clean.

And, of course—tastes better . . . protected from light . . . no deposits . . . no empties to return . . . cools quicker . . . takes up less space . . . no danger of breakage . . . sanitary—used once, thrown away . . . holds 12 fluid ounces, same as a bottle.

Continental Can Company

NEW YORK · CHICAGO · SAN FRANCISCO

THE CONETOP STORY isn't just anti-bottle; it's anti-flattop can, too. The ad from a 1936 issue of *Sports Afield* says that conetops open and pour like bottles ("no special opener required"). Furthermore, you can "drink right from the clean protected top." Notice photo in upper right corner of the ad. See the nice man throwing his empty into the lake? Times change.

In any case, the bottle companies did not take the early success of the beer can lying down. They battled back with a short, no-deposit no-return bottle known as the stubby. And they kept up a steady word-of-mouth propaganda campaign to remind America that a can was not a fit place for beer. For pure, from-the-brewery flavor, they made it quite clear that you should purchase your beer only in glass bottles.

America listened. And the canmakers' prediction that cans would pretty much take over the beer business did not come to pass. By 1941, they had only been able to capture 10% of the packaged beer market.

World War II put a temporary end to the bottle can battle. No beer cans for civilians were produced after May 31, 1942. When the war ended, however, the container industry began where it left off. By 1953, the can companies were able to announce a minor victory: A quarter of America's beer (not counting draft) now came in cans.

The next year saw a newcomer make its appearance on the scene: The 16-ounce can. The first one, according to most historians of the subject, bore the name of Schlitz.*

The 16-ounce can didn't make much of a dent in bottle sales. Neither did the all-aluminum can, first introduced in the late fifties by Coors.

Then, in 1962, a real innovation appeared in the beer coolers of America. The pop-top, pull-tab, or ring-pull (as it was later called).

Iron City was the first beer to offer this easier way to get at a beer can's contents. First introduced in 1962, it meant the beginning of the end for churchkeys—and it gave the metal can an advantage over bottles that wasn't matched for many years to come.

Finally, in 1969, canned beer outsold bottled beer for the first time.

---

*This point is disputed. Some long-time collectors can produce a 16-ounce Krueger can bearing a 1941 copyright. In any case, the first 16-ounce beer to make a big splash on the marketplace was the '54 half-quart Schlitz.

The battle had taken 34 years, but it was now won. In 1970, an appreciative group of St. Louis enthusiasts formed The Beer Can Collectors of America. Not only had the beer can overtaken its arch enemy, the beer bottle; now it even had an organized group of consumers to appreciate and study it.

Sadly enough, few breweries have kept very good records or samples of their can design changes. Not until the beer can collectors got going did a body of information on the subject begin to build up.

Looking back, students of beer cans noticed an interesting aspect of the beer can's evolution. When first introduced, most beer cans bore the word, BEER, in very large letters. They had to, for people were not used to having this particular beverage offered to them in that particular container. Over the years, the generic name for the product has been given less and less emphasis on the label.

Another thing about old beer cans that collectors will forever be grateful for: The older the can, the heavier the paint job on it. The cans from the thirties and forties were much better protected from the elements than today's cans are. As a result, it is not rare to find a can that's suffered three or four decades of abuse but is in much better condition than a can from the early 1960's.

Nevertheless, the days of the beer can—as we now know it—may be numbered. Troubled by litter, a lot of states are making life mighty tough on all one-way (no-deposit no-return) beverage containers.

The long-time animosity of the State of Oregon toward one-way containers is well known. They've banned the pop-top can and forced deposits on all containers. But Oregon's by no means an isolated case. Vermont also has a 5¢ deposit on all beverage containers. And a lot of other states are following suit.

If laws don't do away with the beer can, technology might. Self-decomposing beer cans are the goal of scientists all over the world. It may be only a matter of a few

WHEN BEER AND ALE CANS were a new development, their labels often shouted out "Beer" and "Ale" in type that was even larger than that used for the brand names.

years before a biodegradable beer can will be developed that is guaranteed to perish within just a few days of exposure to sun and rain.

Already, one Swedish brewery has brought out the Rigello plastic beer can. Made of a thin plastic cone (to hold the beer) and some cardboard tubing (to give it

THE BEER CAN as we know it today may be doomed. This plastic Carling can has been tested in England as a possible replacement. It is made of a special resin developed by a subsidiary of Standard Oil Co. of Ohio.

strength), this "can" won't last long along a roadside or in a dump.

According to the English journalist (and beer can collector), W. P. Jaspert, two British beers are also being test-marketed (with some secrecy) in plastic containers with metal pull-tops. They are Worthington and Carling Black Label. Unfortunately, Jaspert reports, these "cans" cannot take high quality printing, so they're quite unattractive.

It's sad, but true. The collector of the future may have to settle for ugly cans. Or no cans at all.

3

Everybody thinks he's the only beer can collector in the world (at first)

> "The day I graduated from college, I found myself without any money, no job in sight, and the rent many days overdue. So, I bought myself a beer. After it was empty, I put it on the shelf."
>
> Martin Landey, Beer Can Collector
> Belmont, Massachusetts

Most beer can collectors didn't start out collecting beer cans.

The beer cans started collecting *them*.

Example: A father was making miniature billboards for his kids' electric train out of beer cans. (He—gasp—flattened them out.) Pretty soon, his friends started bringing him beer cans from all over the country. Every time they took a trip, he wound up with more beer cans. It got to the point where the model railroad was being overrun by beer billboards. So he started saving the cans.

Example: A man's business caused him to move to New York City. He was homesick for Coors, the beer he drank when he lived out west, so he put up a little display of Coors cans in all sizes on top of his refrigerator. Pretty soon, the same thing happened that happened to the father with the model train: Friends started bringing additions to his "collection." He couldn't throw away the unwanted cans—it might hurt their feelings—so he kept them. Pretty soon his "collection" was, indeed, a collection.

Example: A worker at the now-defunct Maier Brewery in Los Angeles, amazed at the number of different brand names they sold the very same beer under, decided to take home one can of each just as a point of interest. Just as a point of interest, that former brewery worker, Ken Jerue, is now one of the largest beer can collectors in the world—and is now working on his *second* collection. (More about that later.)

A lot of collectors kind of slipped—and sipped—their way into the hobby while on vacation. Ben Bright of St. Charles, Missouri is a typical case. He got started when he decided to save a souvenir can of each brand he consumed on a cross-country trip.

Whether they were on vacation when they started tucking away empty beer cans or not, nearly all beer can collectors have committed the very same grievous error when they began: After buying a sixpack of a strange beer they encountered, they saved just one can, disposing of the other five which would've made great traders.*

One novice collector committed an even *more* grievous error. Whenever he encountered a newer version of a can he already had in his collection, *he threw out the older one!*

One of the odder how-I-got-started stories comes from Hamilton, Ohio. A secretary, Louise Durbin, saw an article in a magazine telling of a patio table you can build with beer cans. The plan called for 47 cans, so Louise decided to see if she could find 47 *different* beer cans. She could. But just as she was about to make her table out of them she read about an organization called the Beer Can Collectors of America (BCCA) and the strange hobby to which it is dedicated. "Right then," she says, "I forgot all

---

*The duplicate cans a collector has available for trade are known as traders. A collector who trades cans is *also* known as a trader.

about the table." Today, she has enough cans to build dozens of tables, but would she sacrifice 47 of them? "No way."

As beer can magnate Bob McClure says, "Beer can collecting was like a sleeping giant until the BCCA came along and got him up and moving."

How the BCCA came to be is an amazing story in itself.

On October 20, 1969, an article on the 518-can collection of a St. Louis man appeared in the *St. Louis Globe-Democrat.* The paper thought it was a unique story. (After all, who ever thought of collecting beer cans?) The man, a 56 year old manufacturer of advertising specialties named Denver Wright, Jr., thought his story was unique, too. He assumed he was *the only beer can collector in the world.* A real screwball. One-of-a-kind.

So—as it turned out—did all the other beer can collectors who read the story. Each thought *he* was the only one.

Now that they knew there was another, though, a sixpack of St. Louis collectors came out of the woodwork and gave Wright a call. One of them was Bob Eckert. He told Denver that he had been collecting beer cans for 16 years and had 608 of them arranged alphabetically on wooden shelves in his basement. Another caller was Denver's very own younger brother, Larry. He had 361 cans—and no idea his brother was into beer can collecting, too. The other callers and the number of cans they had in their collections at the time were Ray White (254), Tony Bruning (160), and Kenneth Fanger (140).

These six men began visiting each other's collections, swapping a few cans, and doing a whole lot of talking about the strange hobby they had in common.

What happened next is best described by the Temple University newspaper in a story about beer can collecting.

> The collectors began asking themselves, 'How many others in this great nation have also thought that they alone could see

the true essence of beauty in a can of Olde Frothingslosh?' The six became convinced that they had been chosen by fate to become the groundbreakers in a promising new American Frontier. The die was cast. Each loose thread came together to weave the fabric of a great inevitable destiny. On April 15, 1970, the Beer Can Collectors of America was created in Denver Wright's living room.

After about five months, a whopping total of eleven new members had been added. A year later, in September of 1971, the BCCA could boast 274. In September of 1972, the membership passed 600. In September of 1973, it hit 1,659. By September of 1974, a grand total of 3,215 had publicly admitted their fondness for cylindrical beer enclosures. And September of 1975 found 6,418 people carrying BCCA membership cards. Membership of 10,000 would come during the nation's 200th birthday year.

Each member receives a directory of all other BCAA members, in which each name is fitted with a set of asterisks:

* means the collector has 100-249 cans.
** means the collector has 250-499 cans.
*** means the collector has 500-749 cans.
**** means the collector has 750-999 cans.
*****means the collector has 1000 cans or more.

This enables one collector to properly gauge whether or not another collector would be a likely candidate to receive his or her trading list. A beginner, for example, probably wouldn't have any traders a four-asterisk doesn't already own. But another no-asterisk or one-asterisk collector would probably be interested in a beginner's traders—and vice versa. Beginners need not be worried, though. Col. Clarence Reed (USAF Ret.) of Satellite Beach, Florida went from a collection of almost nothing to four asterisks in just eight months!

The BCCA also supplies a list of all members, broken down by state and city so collectors in a given area can contact each other. The listing also helps collectors when they travel—because beer can collectors are prone to pack a few cases of traders to take along. A phone call in the new location will probably lead to some good trades.

The importance of having allies throughout the country cannot be exaggerated. When a local brewery comes out with a new can, for example, collectors from that area drop notes telling others of the new find, and soon their mailboxes are full of let's-trade-by-mail offers.

In addition to the directory of collectors, BCCA members also receive a continually updated composite list of all the beer cans—both domestic and foreign—known to exist, including individual cans as well as sets.

Every other month, the *Beer Can Collectors News Report* is published, featuring information about new brands, label changes, reproductions of old beer can ads, beer can history, general beer lore, letters from members, biographies of collectors, information about brewery closings, and how-to articles about restoring those rusty old cans you found in the dump last week.

For by-mail trading, the BCCA's *Want Ad Bulletin* is published monthly and consists of nothing but ads like: *Want to trade currents for currents? Like to swap obsoletes for obsoletes? Interested in exchanging a Chief Oshkosh for an Eastside? Etc.**

The BCCA has well over fifty local chapters, ranging from the Badger Bunch in Wisconsin to the Yankee Chapter in Massachusetts. Each chapter holds a trading session every few months to give its members an opportunity for face-to-face trading. Some zealots like St. Louis collector Henry Herbst just can't get enough home-town trading, so he journeys great distances to make the chapter meeting circuit. And never come home empty-handed. Once a year, National *Can*ventions are held. (More about them, later.)

---

*To make sure nobody has a trading advantage over anybody else, the *Want Ad Bulletin* is mailed in stages, going to the collectors who live in remote areas first. As a result, nearly all BCCAers get it pretty much at the same time.

ORTLIEBS BICENT. OR U.S. CURRENTS....SCOTT OTTO #6639, 2311 W. RANCH R
E NEW GIBBONS FAMOUS AMERICAN BREWMAKERS COLLECTION CANS. TRADE FOR AN
FOR CAN....JOHN BOWSER #7556, IVES TRAILER PARK, MADISONVILLE, PA 1844
MEMBER--I HAVE EAST COAST CANS FOR TRADE. ORTLIEBS COLLECTOR SERIES CA
S + MANY MORE. SEND SASE....JOHN PATTON #7638, BOX 221 PASSER RD, COOPE
E ALTES, BALLANTINE RED LETTERING, OLD MILWAUKEE TIN, PABST 7OZ AND MORE
E JONES #7887, 1239 MANCHESTER DR, S. BEND, IN 46615
COLLECTOR WANTS HELP: HAVE TO TRADE OLD CHICAGO DARK, TALL BOY, KOEHLE
ER & OLD GERMAN LAGER & OTHERS....GENE ANASTASIO, JR. #7830, 811 CONODO
FROTHINGSLOSH SILVER & PURPLE, IRON CITY, GENNESSEE CREAM ALE. NEED
H BLICE #7793, 6107 GREAT DANE DR, BETHEL PARK, PA 15102
ED: BLACK LABEL (BOSTON BRUINS STANLEY CUP CHAMPIONS 69-70 & 71-72)..
OBLINGER #7529, 2615 HILLEGOS RD, FT. WAYNE, IN 46808
GAMBRINUS GOLD BEER, PURPLE OLDE FROTHINGSLOSH & OTHERS. SEND LIST.
AN REYNOLDS #6500, 337 PITTSFIELD DR, WORTHINGTON, OH 43085
ED: JUST STARTING NEED YOUR LOCALS (12OZ TAB ONLY) FOR MY LOCALS OR
. SCHOEN #7254, 2113 SYKESVILLE RD, WESTMINSTER, MD 21157
COLLECTOR NEEDS CANS. SEND 6 OR 12 OF YOUR LOCALS, I'LL SEND SAME AMOU
ASELEWSKI, JR #7286, 119 CECILA DR, MICHIGAN CENTER, MI 49254
TAVERN PALE DRY, MUNICH HEIDELBERG & RUSTY PAUL BUNYAN, & SCHMIDTS.
CROWLEY #7512, 3826 S. 57TH AVE, CICERO, IL 60650
D NEW MEMBER WOULD LIKE TO TRADE FOR CANS THAT HAVE COME OUT IN SERI
....DUANE ESSER #7475, ALMA CENTER, WI 54611

THE *WANT AD BULLETIN* of the Beer Can Collectors of America is a good place to advertise the cans you have for trade—and the cans you're looking for.

And one more word on the BCCA: It believes that the fun in building a beer can collection is in trading. As a result, the BCCA takes a dim view of collectors who trade cans for money or vice versa. So the buying and selling of cans amongst members is strongly discouraged.

For a membership application, write:

Beer Can Collectors of America
7500 Devonshire
St. Louis, MO 63119

Another group, the Eastern Coast Breweriana Association, was also founded in 1970, but its point of view is somewhat different from that of the BCCA.

While the ECBA welcomes collectors of all types of brewerians, it was founded for the *multi-interest* breweriana collector, the one who is interested in beer cans or beer trays, tap knobs, mugs, openers, coasters, advertising signs, labels, bottles, bottle caps, or any two or three of the above.

As ECBA President, Bob Gottschalk says, "Where a national organization exists for the collecting of a particular beer item, the ECBA has no desire to compete with them. BCCA is a good example of this for can collectors. I recommend it very highly to collectors who only collect cans."

"At the same time," he adds, "I hope that when a single item collector expands his or her interest to include additional breweriana items the ECBA may then serve them."

ECBA membership runs around 300. Benefits include a listing of all members and the type of breweriana collected by each, a quarterly newsletter, and regular swap sessions and conventions. (ECBA conventions have always been held in cooperation with small breweries.)

For a membership application, write:

E.C.B.A. Membership Secretary
Nellie Winterfield
961 Clintonville Road
Wallingford, Conn. 06492

4

Deciding what to collect: one-sixth of the fun

"The ever-changing designs of beer cans drive me crazy—but I love it!"

Elmer Mick, Beer Can Collector
St. Louis, Missouri

People really go for beer cans in Southeast Asia. They flatten them out and use them for shingles. They also fashion them into pots and pans—even truck bodies.

Nobody's perfect. But to many an aficionado of the beer can, it's a horrendous crime against nature. Which just goes to show that even the jovial, devil-may-care beer can collector can be awfully narrow-minded at times.

After all, each and every collector has to decide for his or her very own self what he or she will or will not collect.

Sure, you can collect every different beer can you can get your hands on. But if you decide to go this route, as many collectors have, you may find yourself in desperate need of another place to live—one home for you and one for your collection.

That's why a lot of can collectors decide to put limits on their collections. Many collect *only* 12-ounce cans. Others collect only 12-ouncers unless they can't find a brand in 12—then they make an exception. Some collect only 16-ouncers and some only want aluminum cans, foreign cans or cans from a single state. Some only

ANIMALS have provided names to nearly as many beers as they have to automobiles. Some collectors specialize in brews that bear the names of wild (and domestic) creatures.

collect cans that start with one letter of the alphabet. Some search for cans with pictures of animals on them, like Jaguar or Big Cat or Mustang. Some choose cans bearing people's names, like Metz or Heileman or Griesedieck. Some want cans that once contained Bock beer or ale.

A number of collectors will only save cans *they've personally emptied.* This makes trading pretty difficult—they can only trade for full cans that they can then drink—but they wind up with a certain closeness to their collections less picky mortals just can't match.

Gil Brennell, former president of the BCCA, has a different set of standards. He's limited his collection to the oldest can he can find of any given brand. He may have a rare old Blatz on his shelf, but if he trades for an even older one, he'll pull the other one out of his collection and put it up for trade.

New Jersey's Bill Christensen specializes in cans from his home state—but doesn't limit himself to them.

Colorado's Ron Carter, on the other hand, has zeroed in on just one brand: Coors. Ron trades for any good can he can get, but only if it will help him get a Coors can he doesn't have. His trading stock, then, includes many brands—but his collection contains just one. Unlike many brewers, Coors has made relatively few design changes over the years, but Carter has over twenty of them. A couple more cans and his collection will be complete!

Compared to Dick Veit's collection, however, Carter's is a biggie. Veit, you see, is attempting to amass The World's Smallest Beer Can Collection. Although he's attended several of the BCCA Canventions, he works hard to keep his collection pared down to sixpack size. Often, this requires 3-for-1 trades, of course, "I'm after quality," says Veit, "not quantity." Others suspect he just likes hanging around trading sessions helping more serious collectors turn full cans into traders.

Most collectors look on the Carters and Veits with a weirder-than-thou attitude. (Or is it the other way around?) There is a real problem, though: How big must the difference in a beer can be to make it a different beer can? These, then are the questions a collector has to answer:

> Will I regard different tops as being different cans? (There have been several varieties of tops on the market: Old-fashioned pull tabs. Modern ring pulls. And now, on some brands, push-in button tops or "two-holers.")

ONE YEAR. That's all it took for Ron Carter to build this collection of over two dozen Coors cans—all different. The three in the second row, far right, don't have the word "Banquet" on them which makes them the rarest in Ron's unusual collection.

Coors
Coors
Coors
Coors
Coors
Light Beer
Light Beer
Fine Light Beer
Banquet

POP-TOPS, pull-tabs, ring-pulls, button-tops, poke-tops, slide-tops, two-holers, flanges. Can manufacturers have experimented with a wide variety of ways for you to get to the contents of a beer can. Some collectors try to see how many different ones they can find. Some don't.

BUDWEISER
PLEASE DON'T LITTER

OVER THE YEARS, cans have been manufactured with a variety of side seams, including soldered, welded, "narrow," "perfect," and none-at-all. Some collectors collect all these variations. Some don't.

Will I differentiate between different metals? (The very same design may be available both in steel and aluminum.)

Will I collect different types of seams? (There are soldered seams and welded seams. And there are narrow seams and wide seams.)

Will I save otherwise identical cans if they're made by different can companies? (Most cans carry the symbol of the company that made it.)

Will I consider a can to be different if it is a slightly different shape? (Some cans have straight sides, others have a crimped edge where the side of the can meets the top.)

Will I judge a can to be a different can strictly on the basis of its manufacturing process? (A brand may switch from the common three-piece can to an extruded* type—also known as drawn and ironed—with no other design change.)

---

*An extruded can is one whose body has been pressed from a single block or "slug" of metal.

SOME STATES require that all beer cans sold there carry tax stamps. Some collectors save all these variations. Some don't.

RECENTLY INTRODUCED onto the market—all five of these cans should be easy for any collector to obtain either at his friendly beer store or from another friendly collector across the nation.

Other minute changes can be detected by the close examination of two otherwise identical cans. A brand may go from one brewing company to another and the only clue will be in the tiny type that tells the brewery of origin. A multi-brewery company may add another brewery to its listing. An older can may lack a zip code after its address, but a newer can may carry it. A can from some states (example: Oklahoma) will bear a tax stamp whereas another can won't. A new run of a can may have colors of a slightly different hue. A newer version of a can may carry the words, "12 fl. oz." at the top of the label although the can that preceded it carried the same words at the bottom.

Just to make things a little more interesting, a couple of developments are afoot that are driving collectors (and their collections) up the wall.

For one thing, the Feds are making all brewers list the ingredients of their beers on all of their labels. (Effective date of law: early in 1977.) A little matter? Well, it makes all the cans of 1976 obsolete and has collectors of minute label changes scurrying around trying to make sure they have both the old and new versions of all current brands. Of course, since they have to alter their printing anyway, many brewers are using this opportunity to make major can design changes.

As the metric system takes hold, of course, there will be another complete set of label changes. These are more likely to be minor ones. The question, of course, is whether or not to consider the addition of "350 ml." as a New Can.

This is something every collector has to decide for himself.

Longtime collector John Vetter draws the line at all minor changes. Period.

Tom Jez says, "A different can is one where the difference can be readily detected by a noncollector."

Bob Eckert (former BCCA president) has well over a dozen different Falstaff cans, yet each is quickly distinguishable from the rest. "When they add a city in small type because they've opened a new brewery or changed from steel to aluminum or made the seam narower or added a tax stamp...nope, that's not a difference to me."

Vaughn Amstutz is known as Mr. Brand Change among collectors. For he is the man in charge of compiling a list of all the little label variations for the BCCA as they come out. Though he's a veritable repository of information about these minute differences, he doesn't collect them. He says, "I have to be able to spot the difference across the room for it to be a new can for my collection." For Denver Wright, Jr., it's a bit more precise. His rule is that a can must appear significantly different to the naked eye at a distance of eight feet. Even more picky is Bob Stutzman, who insists that the change must be perceptible from *four* feet.

Across the room? Eight feet? Four feet? Or at arm's length? Al Stroner, a systems analyst for U.S. Steel, claims to have the most scientific way of all of telling if a can is different. "I put the can in question up on the washer and have my wife look at it. *She* decides if it's different."

What to collect? Every tiny difference in a can? If you do, you'll wind up with four asterisks in no time. But there's a lot to be said for going for quality instead of you-know-what. Including the fact that you'll still have someplace to live.

The problem of what to collect continues to rage, but St. Louis' Bob Feldwisch may have the best solution of all: "I collect any can that I feel like collecting."

Denver Wright, Jr., felt like collecting full beer cans. It's sort of like saving uncancelled stamps. As Canadian collector Ian Russell said, "Full cans have more integrity." Today, Wright owns the world's largest collection of full beer cans. And finding old beer cans that have never been emptied is no easy trick. You won't come across them in dumps. Or along the roadside. About the only

DENVER WRIGHT, JR., and the World's Largest Collection of Full Cans of Beer. *Photo Credit: St. Louis Globe-Democrat*

TED STOLINSKI has over 2,600 cans in his collection—plus these homemade weirdo cans he traded for, just as novelty items. As fellow collector, Bruce Nist says, "I'm sure glad Iron City doesn't really make those little ones. It would take 30 or 40 of 'em just to wet your whistle!"

place you'll run across an old uncancelled beer can is down in the basement of a bar or *waaaay* back in the cooler of a liquor store that hasn't had its stock rotated for a couple of decades.

Nevertheless, Denver Wright, Jr., claims 1,320 different full cans of beer. And that number is not likely to change. Because that's all the specially-made, illuminated shelves in his basement will hold. Behind glass. And under lock and key.

Why the security measures? Because Denver realizes that "it is always possible that some guest would open a can and drink it."

"I haven't increased the quantity for a long time," says Wright. "I've just been improving the breed by making the cans as different as possible. I got rid of misprints, slight variations and the near-beers. I never did collect variations of metal, tops or tax stamps."

There you have it. 1,320 beer cans. All full. But getting better all the time.

And lighter, too. "Some of them must be evaporating through invisible pinholes!" Evaporation, of course, isn't to be feared like leakage is. Imagine having 12 ounces of ancient rotgut dripping down your shelves!

STRANGE THINGS happen to beer can collectors after they have several thousand cans in their collections. Clay Tichelar is a case in point. In his spare time, he paints garbage cans.

Fortunately, Wright hasn't had many problems with leakage.

One or another of the 1,320 cans will spring a leak on an average of every two or three months, but that's all.

Wright attributes his good fortune in this regard to his hands-off policy. "Don't bother the cans." he says, "and they will behave."

The reason Denver Wright collects full beer cans, of course, is because he *started* collecting full beer cans. There's a lesson here for all of us.

"Frankly," admits Wright, "I wouldn't recommend that anyone save full cans. I got started that way and couldn't quit. Over the years I've passed up some mighty rare cans because they were empty. Another thing: Full cans are heavy and that makes a difference in the kind of shelving you can put up. Finally, postage is a much bigger expense when you're trading full cans by mail. Of course, there's the question about how legal it is to send full cans through the mail, too..." His voice trails off.

Amazing. The world's largest collector of full cans admits full can collecting is as full of drawbacks as it is of beer.

If you're just starting to collect, best you take a moment to ponder whether the restrictions you've put on your collection are ones you're going to want to live with.

For the next 500 years.

5

How to tell how old a beer can is

"Every time I see the words 'Internal Revenue Tax Paid' on a beer can my heart goes pitter-pat."

Bob Gustafson, Beer Can Collector
Golden, Colorado

The canny can collector likes to know how old his cans are, and there are some pretty good ways to determine it. Or at least get an approximation.

If the can bears the four magic words, "Internal Revenue Tax Paid," for example, you know that is a quarter of a century old and then some. By law, all cans filled before March 1, 1950 had to carry that statement.

Another clue to a can's age is the thickness of its metal. Usually, older cans were made of heavier gauge steel. Of course, if a can is the punch-top type that required a churchkey to open, it probably was produced prior to the mid-1960's when the pull-tab top became popular. (One exception: Because of Oregon's recent law against pull-tab tops, modern-day cans shipped to that state are once again the old fashioned type which collectors know as punch-tops or flattops.)

Although Coors produced some aluminum cans for part of their marketing area in the late 1950's, aluminum cans didn't really make a splash in the marketplace until the mid-1960's. So don't let anybody tell you an aluminum can is a real oldie—it can't be.

If a can bears either BEER or ALE in very large letters, the chances are good it's a pre-World War II can. As mentioned before, the idea of buying brew in cans was still so new to many people back then that can designers felt they should make it very clear to the consumer that there was, indeed, BEER or ALE inside that strange enclosure.

A few brewers stubbornly continued to put their products into conetops as late as the mid-1950's, but most conetops other than Kessler, Falstaff and Grain Belt are much earlier cans.

And any can that has a diagram on it to show the consumer how to use a churchkey is definitely a vintage item. After all, it didn't take long for America to learn how to get at the contents of a beer can.

Dating beer cans, as you can see, is a pretty complex art. It isn't easy to sort out all the clues, evaluate them, and come up with an accurate estimation of when a can started out in life. You have to be a detective.

This being the case, Sherlock Holmes' Sudsy Brother is a California banker by the name of Bob Myers.

When it comes to determining when a can was produced, Myers has no rival. He wrote the book on it. Literally. (His booklet, "Some Thoughts...On The Beer Can's First 35 Years," was in fact the first book ever published for beer can collectors.)* Myers' research into the subject is thorough and his knowledge encyclopedic. A sample: "During 1935-1938, American Can Co. used the term 'Keglined' with 'patent pending,' then with 'patents pending.' Early in the 1950's, this material was progressively reduced in size—from a full length panel to just a small block near the seam. By the late fifties just a circled 'Keglined' remained—and by the mid-60's all reference to the lining was discontinued."

*Robert Myers, "Some Thoughts . . . On the Beer Can's First 35 Years" (Santa Barbara, California: Myers, 1971).

KEGLINED was American Can Company's name for its can coating (to keep the beer from coming into contact with the steel). As the years passed, the term became less and less important to the consumer. These three cans, all the same brand, show how the Keglined designation was reduced in prominence in stages. Finally, it was dropped completely.

THREE TYPES OF CONETOPS. The Cremo has what is known as the "J" top made by Crown Cork & Seal. The other two are by Continental Can. (The low-profile Apache, by the way, is a mighty rare can. Only five are known.)

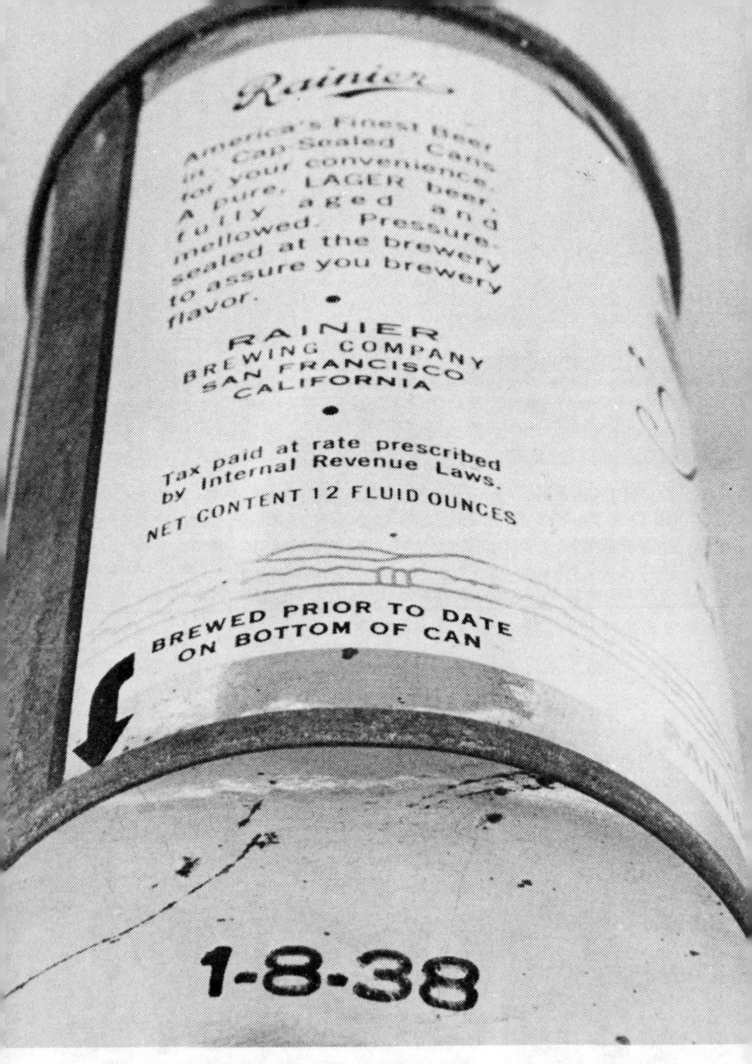

SOME OLD BEER CANS come right out and tell you when they were produced. Which takes all the mystery—and fun—out of trying to date them.

His analysis of the early cans of Continental Can is just as impressive. "Continental, pioneers of the 'spout' can, used the trademark 'Cap-Sealed.' In general, cans that carry this trademark in block letters came out prior to 1938. After that year, it was generally in script. Around 1950, the term was phased out."

There's more. "Very early Continental 'spouts'—the first six months worth—had *flat* bottoms. And, because the spout was unvarnished, it rusted rather quickly. Another quick clue to dating Continental's spouts is the height of the spout. Pre-WWII cans had a lower profile compared to those used during and after the war."

If you've got some cans that were made by a brewery that's no longer in business, of course, you can find out when the brewery closed its doors (and vats) and then you'll know that the can is at least ten years old.

Many collectors spend a lot of time rummaging through old magazines eyeballing beer ads because the can designs pictured therein provide a positive method of establishing when the cans in their collections were current.

Someday, perhaps, a catalog of every beer can ever produced will be published that will include accurate dates-of-issue data for each. It will take a lot of research, of course.

And a lot of Bob Myerses.

6

Some beer cans

"I've never met a beer can I didn't like. Some, of course, I like more than others..."

Bob Leslie, Beer Can Collector
Worthington, Ohio

Beer cans. That's what it's all about. Putting all the tough (hard to find) cans you can find on your display shelves.

If you had every can ever made—around 20,000—and stacked them 6 feet high, your shelves would be over 300 feet long.

Of the 20,000, however, some would stand out as being a little more special than the rest. They wouldn't necessarily be the prettiest cans, nor would they be the rarest. They'd be the most *interesting* cans. (Unfortunately, because of trading pressure, the most interesting cans often develop into the "rarest" cans even though they're not really the rarest in terms of numbers available.)

Let's look at some cans. Some are tough, as the collectors say. (But a tough can is an easy can if you find it under an old building, along a roadside, or way back in the cooler of your neighborhood liquor store.) Some aren't tough at all. But all are tough to beat when it comes to being nice cans to have holding your shelves down.

A NUMBER OF CANS have numbers in their names. Here are 19 such cans.

A FEW YEARS AGO, a couple of brewers thought there was a huge market for fruit-flavored beer. They were wrong.

FOUR CANS OF THE YEAR, as voted by the members of the Beer Can Collectors of America. Except for Colorado Gold Label (by the now-defunct Walter Brewing Co. of Pueblo, Colorado), all are currently on the market.

Let's start with Acme. But not because it's the first in alphabetical order. There are a couple of dozen cans ahead of it—various versions of ABC Beer and Ale and Ace Hi. We start with Acme because it has a history that goes way back into Internal Revenue Tax Paid days—and yet it lives today. No matter that the Acme you find at the supermarket (if you live on the West Coast) or on another collector's trading list is a distant

cousin of the original, it's still an Acme. And its forbearers were pretty outrageous cans. For years, they used to bear this slogan: NON-FATTENING REFRESHMENT. If that semi-nontruth weren't enough, the old Acme cans also used to promote their contents as HEALTHFUL and INVIGORATING, too. It was just a hop and a barley from snake medicine, really. Today's Acme can doesn't sport such ludicrous claims for any alcoholic beverage—but it's still mighty collectible because it's such a beautiful can. (In 1975, it was named The Can of the Year by the BCCA.)

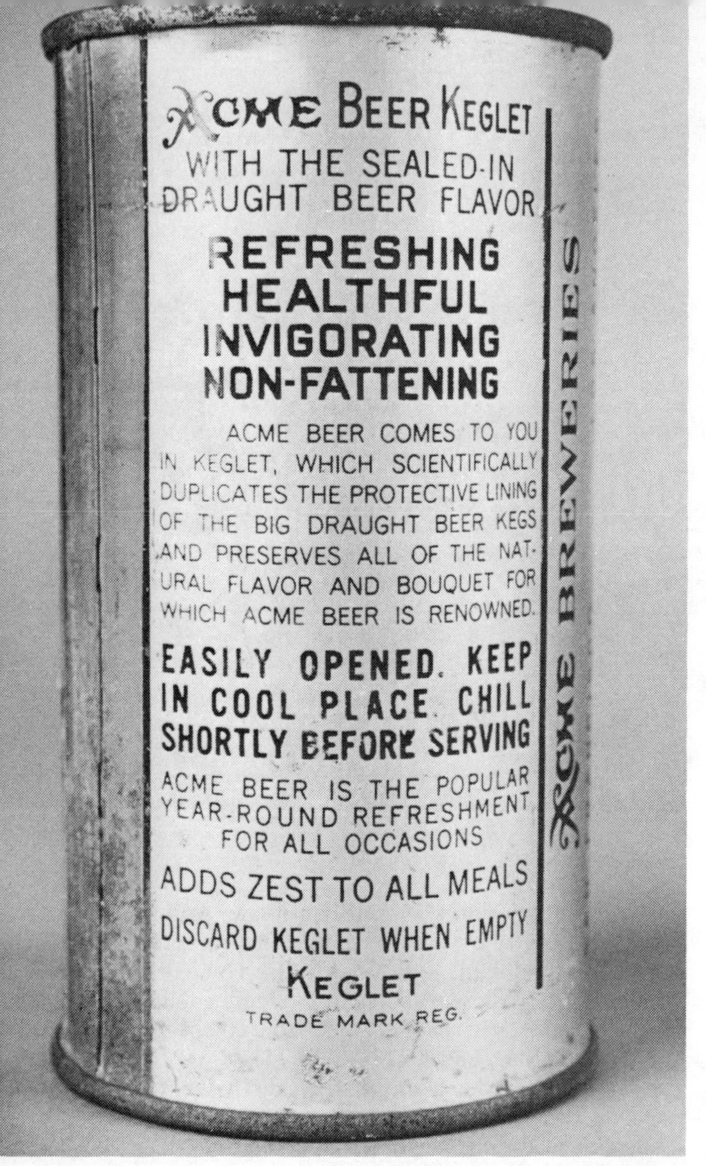

NOT ONLY does the side panel of this old Acme can sell the Keglet (as they called it) as a fine container for beer, it also tries hard to convince the consumer that the beverage inside is both healthy and non-fattening.

VITAMIN-ENRICHED BEER would never be permitted today because the Feds don't approve of making health claims for alcoholic beverages. But back in the old days. . . .

BEER BELLIES BEWARE! A number of low-cal brews have been produced over the years. Here, complete with legal disclaimers (". . .than our regular beer") are two of them.

When it comes to making blatant health claims, though, no one can match the famous Schlitz Sunshine Vitamin D can. "This can contains 100 U.S.P.X. Units of VITAMIN D, The Sunshine Vitamin," it said. You can just imagine the beerdrinkers of back then slugging them down "just for health's sake."

Beer's reputation as a calorie-laden beverage, of course, has always been a problem for the brewing industry. No wonder that there have been a number of brands that have claimed to leave you just as svelte after a few sixpacks as before. Mark V, Gablinger's and Dawson's Calorie Controlled Ale have all promoted themselves that way.

Today, this concept continues to be a vital one to the beer marketers of America. Schlitz Light and Miller's Lite are carving out huge chunks of the market with their lower calorie brews. In a recent year, for example, Miller's sales shot up over 40% and the gain can be attributed almost entirely to the success of Lite.

Storz-ette, an Omaha brew, wasn't just low-cal, it was aimed right at the lady beerdrinker. Available only in 8-ounce cans, Storz-ette was presented as a beer strictly for females. Which is a real rarity in the history of the beer industry.

Male-oriented beers have always been much more common. Like Mr. Lager. To the total amazement of the women's libbers of the time, it had the effrontery to boast that it's "For Men Only." Genesee was only slightly less blatant when they came out with "The Male Ale."

SEXIST CANS. Storz-ette ("The Original Beer for Women") is flanked by Mr. Lager ("For Men Only") and Genesee Cream Ale ("The Male Ale").

IN 1956, RHEINGOLD put all the contestants for Miss Rheingold of 1957 on separate cans—and then gave the winner another special can. Playmate was another limited-run girlie can. Playboy magazine put it out of business through legal action. James Bond's 007 didn't have legal problems, just sales problems. So it, too, was only made in limited quantities. Olde Frothingslosh, on the other hand, is currently being produced in abundant quantities.

If Mr. Lager and The Male Ale were sexist, there have been a lot of other brews that leave both far behind when it comes to making a broad appeal (!) to the male beer-drinker.

One of the better known of the girlie cans is the Miss Rheingold series. Although the Miss R. contest ran for

many years, the contestants were pictured—all six of them—on Rheingold cans only in 1956.

If the Rheingold girls are desirable, the James Bond girls are lusted after. The official name of the product inside this series of cans is James Bond's 007 Special Blend. It was supposed to be beer plus malt liquor. But what was inside the cans means very little to collectors compared to what was outside. There were seven cans—each featured a different girl—and not many of these cans were produced. National Brewing gave James Bond and his gals a six month test in 1970 in just five cities. Only 18,000 cases of cans were issued. By any

LAGER LOVELIES is what Scotland's Tennent Caledonian Breweries calls one of their five sets of pin-up cans. Here's a sample can from each. (Some are available in some U.S. liquor stores.)

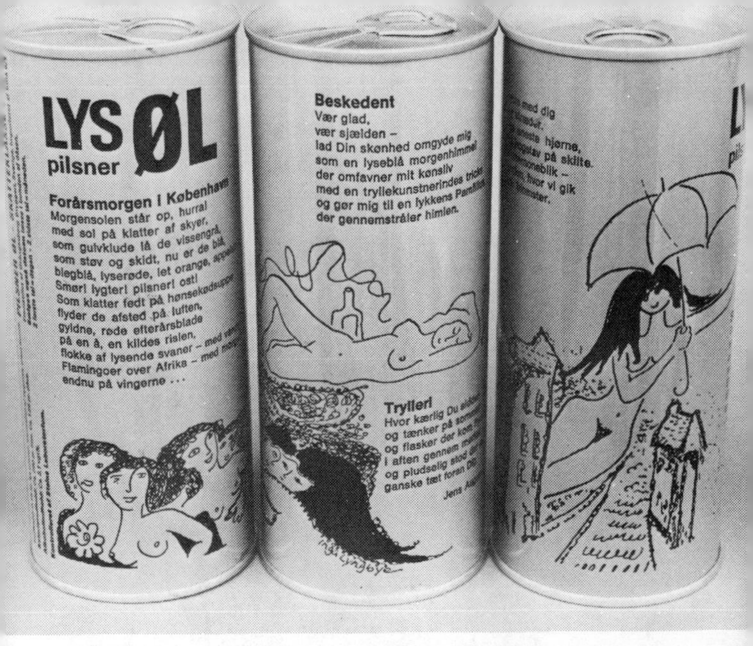

THE SO-CALLED PORNO CANS from Sweden bear racy poems and sketches. They're currently on the market—but not in the U.S.

standard that isn't much beer. By the standards of the beer can collector it wasn't nearly enough. Tough cans? Yup.

Another tough lady is Playmate, offered to the public by the Sunshine Brewing Company of Reading, Pa. But not for long. Because Hugh Hefner's organization successfully sued to get it off the market.

U.S. brewers aren't the only ones who've gotten a gleam in their eyes and tried to use sex to sell beer. Their cousins across the ocean have, too. And have succeeded where America failed.

From Scotland comes Tennent's Lager. You'll have to drink quite a few sixpacks to add all the Tennent's pinups to your collection, though. There are several series of them, all featuring full-color photography, and the pictures on the standard size cans are different from those

SPORTY CANS have long been a favorite of collectors. The Tiger can is particularly interesting. It was brought out (for a short time) by Jackson brewing after the 1958 edition of the L.S.U. Tigers went 10-0 for the season.

SPORTS COMMEMORATIVES, one Carling (for the Boston Bruins Stanley Cup victory) and three from Pittsburgh Brewing.

Contents 12 Fluid Ounces
BOSTON BRUINS
STANLEY
CUP
CHAMPIONS
1969-1970

MOUNTAINEERS
NO. 1

Congratulations
Super Steelers
1975
PITTSBURGH 16
MINNESOTA 6

IRON CITY
BEER
IRON CITY
BEER
The Steelers, 1975

Fehr's
THE BEER
X
L
THAT XCELS

RANK FEHR BREWING CO., INCORPORATED. LOUISVILLE, KENTUCKY.

BEERDRINKERS can keep tabs on their favorite teams if they have one of the schedule cans handy. (The Colt 45, you'll note, is a bank can—a promotional can in which you can save your spare change.)

CROWNTAINERS were the "all-steel bottles" made by Crown Cork & Seal until the early 1950's. This one, from Louisville's Fehr Brewing, depicts the Kentucky Derby.

on the larger cans. (And, as many students of such things have observed, the bigger Tennent's cans are a bit spicier than their little sisters.)

Many of these cans are currently on the market, and most collectors are eager to get their hands on the Lager Lovelies (as Tennent's calls one series of its girlie cans).

A bit tougher to obtain (because, unlike Tennent's, they're not imported to the U.S.) is the four can set of so-called porno cans called Lys Ol from Grangesberg, Sweden. Are they really racy? Well, the line drawings on these cans do show ladies sans clothes. And there are some limericks. But they're not in English, so who knows? Well, actually, thousands of cases of this brew have already been sold in Denmark, so *somebody* knows.

Speaking of sports, many's the brewer who has tried to sell his beer by putting it in sports-oriented cans. A classic example is Cook's 500 Ale, a can designed to hype sales of the brew in Indiana during the Indy 500.

Other sporty cans include Dodger Lager (supposedly only made for distribution at Dodger Stadium), Twins Lager (made in Duluth, it just might have something to do with a baseball team located nearby) and Home Run (a powerful brew sold only in Puerto Rico where baseball is *very* big). Then there's Louisville's pride (of many years ago)—Fehr's X/L which depicts racehorses all the way around the can. Some Louisvillians say this can was only produced during Derby Week.

In terms of numbers produced, these one-short cans can't hold a candle (or a bat or a football) to the numerous Iron City cans which have carried the schedules, rosters and win-loss records of nearly every team that has ever taken the field to defend the honor of Pennsylvania. Other love-our-local-team cans from around the nation include Colt 45 (with the Golden State Warriors schedule), Gambrinus (with the Ohio State success story), two Carling Black Labels (with the Boston Bruins' Stanley Cup victories celebrated thereon), and A-1 (with the Phoenix Suns' and Roadrunners' schedules).

THREE BRANDS of beer that were aimed at a specific ethnic group—Blacks. Peoples was produced by a brewery owned by Blacks. Soul is the legendary brew that was supposed to have been sold only during the Watts disturbance—but wasn't.

If aiming your cans right at the sportslover is a natural for the brewers of America, so is aiming them right at an ethnic group. After all, many Americans claim to be Giants fans (or Dodgers or Yankees or Pirates) first—and Italians (or Irish or Germans or English) a close second.

As a result, many beer cans have carried names heavy with ethnic intonations, Bavarian, Bohemian, Fitzgerald, Holland, Irish Green, Jolly Scot Ale, London Bobby, Old Dutch, Salzburg, Tam O'Shanter, Viking, and Old Vienna.

Some of the most interesting cans, however, have come from breweries that have tried to make inroads specifically among Black people. Black Pride was one. People's (a brewery owned by Blacks) was another. And the most famous of all—Soul.

AL CAPONE'S BREWERY, Canadian Ace, was responsible for this brand.

MONTANA was once the home of several breweries—now there are none. The brewers may be long-gone, but their cans remain—on display in collections from coast to coast.

If you asked 100 average beer can collectors what can they'd most like to add to their collections, 75% would probably say Soul. (The rest would say the first Krueger can.) What makes Soul sc special? "Well, it was only made during the Watts riots," they'll say. But they're wrong.

The truth is that Maier Brewing distributed Soul during 1967, two years *after* the riots were over. Why, then, is it so rare? Because collector pressure has made it rare. Intrinsically, it has very little going for it. Maier has made many a brand in its time and this is just one of the many. Kenny Jerue (who worked for Maier) poo-poos Soul. "There are lots of cans I'd rather have than Soul. And lots of cans I'd trade Soul for. It's not that tough a can."

Jerue would get a pretty good argument out of all the collectors who haven't yet been able to get their hands on one of the Soul cans,* but he's probably fairly accurate in his appraisal of the Soul situation. Soul probably *isn't* that rare. There must be 1,000 of them in collections right now. You can't say that about Apache. Or Clipper. Or Ranger. Or a lot of other cans.

Which brings up one of the problems of establishing values on beer cans. The rarest cans of all are the cans that only one or two collectors have. But these cans are not perceived by the collectors at large to be the super-rare cans they actually are. To build a reputation as a very rare can, there have to be enough of them around for collectors to talk about and trade.

So cans that are in such short supply that no active trading goes for them never become well-known as the super, *super* rarities they are.

---

*There are at least three varieties: Soul Mellow Yellow Beer, in 12 and 16 ounce sizes, and Soul Malt Liquor, in 16 ounces only. Most collectors believe the first-mentioned to be the scarcest, the last-mentioned to be the least rare of the three. Whether or not there was a 12 ounce malt liquor version is still being debated.

7

More beer cans

"Oh boy! More beer cans!"

Harry Keithline, Beer Can Collector
St. Louis, Missouri

Sometimes, the name of a beer brand is what makes it special.

Let's face it, Schlitz, Budweiser, Miller and Hamm's do not set fire to the Soul. Or match it in dramatic impact.

But a lot of lesser brands—ones that never provided much of a challenge to The Big Boys—do.

Listen to these names. Let them roll around for a while in your mind. Bulldog. Cloud Nine. Golden Gate. Eureka. Happy Hops. Nu Deal. Paul Bunyan. Simon Pure. Tube City. Wooden Shoe.

Now *there* are some inspired names. None of them comes encumbered with a heavy story behind it, but they all sure stand out from the Ballantines, Coors and Rheingolds on your shelf.

TWO HUGE BRIDGES, the Mackinac and the Golden Gate, have inspired beer can designers. Both (the cans, not the bridges) are now obsolete.

If Eureka and Cloud Nine and Bulldog ("A pip of a nip in every sip") aren't exotic enough for you, of course, you can always stock your shelves with cans with names like Nimmos, Riks, Urpriator Starkbier, Viikinki Vaalea, Hummer, Nastro Azzuro, Tre Hjarten Fatol and Pilforth Biere Blonde.

LITTLE 7-and 8-ounce cans are the favorites of many collectors.

Canned beer in large quantities is a fairly recent development outside North America—bottles dominated take-home consumption for decades—but it's in full swing now. And the opportunities for collectors are tremendous. As printing expert Bob Stutzman says, "Foreign cans are much prettier than domestic cans. Their graphics are stronger, they're more colorful, and the lithography is far superior to most of the cans made over here."

To add to the intrigue of collecting foreign cans, a high percentage of breweries outside the U.S. is now producing cans in sets. The Swedes and Scots seem to be the leaders in this area, but some fine sets have come from the brewers of Japan, too.

The main problem when trading for foreign cans, however, is establishing values for them. If this is difficult to do for domestic cans, it's even harder to do for cans we're not so familiar with. Which foreign cans are current? Which are obsolete? Which are oldies? Which are really scarce? A few collectors like Darrold Russell and Don Kurtz have concentrated on foreign cans in recent years and their studies have paid off for them, but for most collectors the subject remains, well, foreign. This is why most collectors concentrate on domestic brands, picking up foreigns only when they're easy. Or when a friend brings some back from an overseas trip.

One longtime student of beverage container design, Thomas Reynolds of Madison, Wisconsin, offers an explanation for the design superiority of foreign cans.

"Their cans are better looking than our cans primarily because we've been canning beer longer than they have. Most of America's long-time brewers—and they're the ones who have blessed us with the most cans—feel compelled to stick pretty close to their original designs. They

FOREIGN CANS at right are often more colorful and beautiful than domestic cans. There are reasons for this.

SUNTORY BEER
BY SUNTORY

Gränges
Export Beer III

SKEPPAR
ÖL
CURAÇAO · 1825
Netherlands Navy

THE LOWLAND BRIGADE
A PIPE SERGEANT OF THE
ROYAL HIGHLAND FUSILIERS

GOATS are frequently pictured on Bock cans. It's a long-time tradition.

fear that if they redesigned the Schlitz can, say, they'd lose the recognition they've built up for it over all those years. Overseas, cans for beer are relatively new, so they don't have this problem."

Reynolds is probably right. Most major U.S. brewers have been very cautious about tempting the fates with major label revamping. The Pabst of today is very clearly a descendent of the Pabst of long ago. The original Coors can of 1936 could be easily mistaken for the current one. And the same is true of several others.

The best-looking domestic cans on the market today are ones that aren't burdened with a past to live up to. Huber Brewing's Our Beer is a good example. With its beautiful full-color photography, it could easily pass itself off as a foreign entry. But only a few long-time brewers have taken their chances and put their well-known brands in brand new packages. Luck Lager is one of those few.

For the most part, we're stuck with slight variations of old-fashioned designs out of the 1930's. In addition, there are a number of traditions that many U.S. brewers are reluctant to go against. That's why they so often use red or blue lettering for beers, green for ales, and brown for bock. And that's why bock beer labels almost always carry a picture of a goat. Why? Who knows? It's *tradition.*

There was only one time in U.S. history when a few of the more adventuresome brewers *did* make semi-major label changes. That was when they brought back canned beer after World War II. Apparently, they figured folks had forgotten what their beloved beer cans had looked like before the war. Therefore, a number of brewers were emboldened to freshen up their labels a bit. Nevertheless, the point remains: We're stuck with variations of designs created in the 1930's—and the early 1940's.

A lot of marketing wizards would say that's the way it should be. After all, a beer can should *look* like a beer can. And Americans are used to beer cans that look kind of stodgy, old-fashioned and, well, *beery.*

Just because America didn't have beer cans during the war doesn't mean that Americans didn't have them. A number of brands donned camouflage and went overseas, much to the delight of Our Boys. Schaefer was there. So was Fort Pitt. And Hamm's. Carlings was, too. All

DURING WORLD WAR II, several brands of beer donned olive drab and went overseas. Known as camouflage cans, they are all very scarce today.

drably done up in shades of olive so that snipers wouldn't spot our fighting men while they were drinking. (All camouflage cans are now as scarce as nylons were then.)

Then there's the incredible story of the Olde Frothingslosh can. Once it, too, was scarce. But not anymore.

Without meaning to, a Philadelphia disc jockey named Rege Cordic created Olde Frothingslosh. Day after day, he treated his KDKA listening audience to bogus commercials for a nonexistent beer. "Drink Olde Frothingslosh," he would say, "It's the pale stale ale. . .the beer with the foam on the bottom!" And on and on. All for the amusement of his listeners.

But not all were amused. Some took him seriously and began to demand Olde Frothingslosh at bars, in restaurants, and at package stores throughout the area. Others realized it was a gag, but they demanded it, too.

Then, in 1954, the management of Pittsburgh Brewing had some labels printed up and gave 500 cases of bottled "Olde Frothingslosh" to friends of the brewery. Just for fun. Inside the bottles, of course, was Iron City beer, but the recipients of the unique package got such a kick out of it that the brewery decided to offer it to the public the next year—as a special Christmas season treat. Once again, it was in 12 ounce bottles, but now there was also an 8 ounce *canned* version called Sir Lady Frothingslosh. About a quarter of a million of these cans were sold, but only twenty or so are still known to exist.

For the next twelve years, Old Frothingslosh was sold only in bottles and only during the holiday season. Then, in 1968, a 12 ounce Olde Frothingslosh can was introduced. The relatively few beer can collectors that were active at the time snatched them up. And they were glad they did because for quite some time thereafter it appeared that O.F. had made its one and only appearance in a regular-sized can.

Those collectors who had these cans to trade were able to swap each one for almost anything they wanted. Not only was it very rare, it was also a real eyepopper up on the shelf. Olde Frothingslosh, you see, featured a photograph of the behemouth Fatima Yechburgh. Obviously a spoof on the Miss Rheingold contest, the label claims that Fatima was named Miss Olde Frothingslosh because of her "beauty, talent, poise...and quantity."

While Fatima's quantity was rather substantial, the quantity of cans bearing her picture that were available for trade wasn't, so their rareness and desirability continued to grow until *it* happened.

It was Pittsburgh Brewing's decision, in the winter of 1973, to reissue Olde Frothingslosh. As vast numbers of these cans poured out into the marketplace, the value of the supertough can plummeted. Today, the brand is even available in an array of colors. And new ones come out all the time. Meanwhile, collectors who gave an arm and a leg for the original have to take it down from the shelf and check out the top to reassure themselves that they do, indeed, own the original. Because the only difference between it and the current version is that the tops of the

Olde Frothingsloshes made since '73 bear exhortations not to litter whereas the earlier ones didn't. (Unfortunately, not many collectors collect top variations. Alas.)

Another tongue-in-cheek brand met a similar fate. King Snedley's, by name. It all began when Lucky Brewing researched its market (California, mostly) and discovered an alarming fact: Their appeal was weak to the beer-guzzling 18 to 25 year old bracket. Instead of trying to reposition Lucky Lager, they decided to develop a new brand and take direct aim on the younger drinkers of California.

The new brand was King Snedley's, and it *was* new. Instead of just putting their old beer in another can—a common enough practice among brewers—they formulated a new brew that scored very highly with college students in taste tests. The can carried drawings of a fictitious monarch and his retinue. And the laugh-it-up advertising encouraged folks (especially young folks) to "Look for the royal family on the can!" For some reason, Lucky's strategy was deemed a failure and the brand was pulled off the market after less than a year.

As time passed, King Snedley cans became rarer and rarer. They had almost reached the same I'd-give-half-my-collection-and-throw-in-my-car-for-one status that Olde Frothingslosh had when, in 1975, King Snedley reappeared on the scene. Things like that just don't happen with rare coins and stamps.

Suddenly, everybody has a Snedley in his collection and a couple of cases to trade. The value, of course, has now dropped to one-for-one for another current can. And even if the brand is pulled off the market again, all those cans of traders that are tucked away would keep the value low for years to come.

There are, however, a few supertough cans that you can rest assured will never be devaluated by being reissued. Du Bois Budweiser, for example. After a lengthy court battle, Anheuser-Busch finally put the stop to little Du Bois Brewing's use of the famous Budweiser name

ONCE A MIGHTY RARE CAN, King Snedley's is common, again. Issued at first in limited numbers, it has reappeared on the scene in vast quantities.

and forced them to pull Du Bois Budweiser off the market. So Du Bois replaced the offending word with one that nobody—not even the world's largest brewers—could claim sole ownership of: Premium.

As mentioned earlier, the folks at *Playboy* were also successful in their legal skirmish with Sunshine Brewing over the name Playmate.

As Iowa collector Curt Fisher has observed "What happened to Olde Frothingslosh and King Snedley could happen to Soul, 007, or most of the other tough cans. The only ones we know that won't be reissued are Playmate and Du Bois Budweiser."

Another one may be Amana Beer. Produced only in late 1975, this brand barely got a chance to get off the ground before The Amana Society of Iowa, one of America's early experiments in communal living, got an injunction to put it out of business. At least for a while.

Collectors have a recurring dream: having his very own brand of beer made so that he'd have something to trade that nobody else has. Is it out of the question? Not if you don't mind buying several thousand cases of Your Beer; that's the minimum order from most breweries that produce private label cans.

Private label beers—brands that are made just for one retailer rather than for general distributor—have put some mighty nice cans up on the wall. You'll never find these house brands advertised on network TV, but, they make excellent traders when you're negotiating with a collector from across the nation, because they're generally available only from a few local outlets (sometimes only one). Besides, these brands don't have any pretensions about them. It's just good old beer—and it's priced accordingly. As a result, investing in a case or two of the local cheap-o brew probably won't break you. However, if the good old beer is more old than it is good, you may have to pour it down the drain. And, if you're a real beer-lover, that *will* break your heart.

LEGAL BATTLES stalk the unwary brewer. Anheuser-Busch took exception to DuBois' use of the word, "Budweiser," and successfully sued to have it removed from the DuBois Beer can. (It was replaced with "Premium.") A court order stopped production of Amana Beer in late 1975, if only for a while.

PRIVATE LABEL BRANDS are the beers a brewer packages under somebody else's name—usually for a store or chain of stores that wants to carry its own brand of beer.

Beer drinking—like beer can collecting—is a matter of personal taste, though, and there are lots of collectors who'll swear on a stack of Bartenders' Guides that they actually *like* a lot of those private label brews.

Safeway's house brand, Brown Derby Lager, is probably the private label in widest distribution. The unique thing about it is that over the years Safeway has had it produced by over a dozen breweries. At any given time, the Brown Derby you buy in, say, California is really a different beer than the one you encounter in Wisconsin.

In areas where supermarket chains are permitted to sell beer, house brands are quite commonplace. The best known, of course, come from the biggest grocery chains: Bohack, Giant Food, Maid Rite, Schwegmann, Grand Union, Rialto (also for Grand Union), Finast, Shoprite,

DID SOMEBODY try to copy somebody's else's label? Sure looks like it.

Shopwell, Tudor (A&P), Crystal Colorado (Von's), Our (Del Farm and National), Prize (Food Fair) and Hynne (Fed Mart Stores).

In some cities, 7-11 Stores have their 7-11 Beer, too. So do a few drug stores like Dart Drug (Dart Premium) and Skagg's Drugs (Katz Premium). For awhile, Foss Drugs, a one-store operation in Golden, Colorado, offered its own Ski Country Beer.

G.E.M. and G.E.X. Beers have come from the government employees' discount houses of the same name.

A handful of liquor stores have offered their own brands, too. The most frequently seen (at least on trading lists) is 9-0-5, available from the large chain of 9-0-5 liquor stores throughout the midwest.

Other less-common liquor store brands which may or

LOOK-ALIKE CANS. To save money, the brewer probably decided to reuse an old set of plates. There are several pairs of cans in existence that are as similar as Kings' and Hapsburg.

may not be currently available (they come and go) include the following New Jersey brews: Wilco (for Roger Wilco Liquors), Park (for Park Beverages) and both Astro and Astor Home (for Home Liquors). Not to be outdone, Minnesota has also supplied collectors with some interesting liquor store brands, too. There's been Golden Valley, Jennings, and Sternewirth. (Sternewirth, a tough can to find since it's been off the market for several years now, was sold only by Otto's Liquors in Mendota.) Blanchard's is the house brand of the Blanchard's chain of liquor stores in Massachusetts. Denver's huge Harry Hoffman liquors offered Hoffman House beer until its supplier, Walter Brewing of Pueblo, closed down a couple of years ago. Still, a fairly good number of these cans are still kicking around for trade.

When it comes to scarce private label beers, though, there's probably nothing that can match a Roebling conetop. Dated 1957, Roebling is said to have been produced for the 25th anniversary celebration of a New Jersey roadhouse's getting its liquor license back after Repeal. (In the interim, of course, it had operated as a speakeasy.)

We can assume, then, that a lot of collectors will be hanging around that very same roadhouse when their 50th anniversary comes around, just on the off-chance they'll have another commemorative private label conetop.

Better circle 1982 on your calendar.

8

Still more beer cans

"What beer can do I want most of all? That's easy. *Any* beer can I don't have!"

Harold Lorenz, Beer Can Collector
Cedar Falls, Iowa

When a brewer introduces a new beer and it quickly goes bust before it's gotten much distribution, there are generally just two things a collector can do: Go berserk, grab the phone and call his collector pals who live where the beer was distributed.

A number of short-run cans have thrown the hobby into a state of mild hysteria over the years. Legal problems, as have already been mentioned, halted Amana after 1,700 cans were produced. Dunk's (a mid-1975 product of Florida's Duncan Brewing) seems to have met a premature death because of the rising cost of cans. Some say that only about 250 cases of Dunk's were produced, but controversy continues to surround this brand and some collectors question how tough a trader it really is. Late in 1975, Rainier came out with a new ale can. On the back was a detailed statement of all the wonderful things Rainier put into its ale. Unfortunately, the Washington State Liquor Board objected to this statement and the design was quickly discontinued.

Back in 1955, the experiments of Anheuser-Busch dropped some short-run cans on the market that still

ONE OF THE VERY FIRST commemorative cans was Ballantine's special edition to honor the 1939 World's Fair. Shown here is the back of the can.

CLASS REUNIONS have occasionally caused promotion-minded brewers to produce special commemorative cans. Sometimes, they're topless, and just issued as promotional drinking cups.

have collectors shaking their heads. It seems they were toying with the idea of offering a popular-priced partner to Budweiser and Michelob. Finally, they came up with Busch Lager. Reportedly, this brew was Anheuser-Busch's first artificially carbonated product—and it didn't quite make it from the standpoint of taste. When beerdrinkers started complaining, the brewer wisely pulled it off the market before it did irreparable damage to their reputation.

THE BICENTENNIAL brought forth a bonanza of red, white & blue commemorative cans. Here is but a sampling.

TIVOLI BREWING'S LAST GASP. A beer the struggling brewery hoped civic pride would help become a big seller in Denver. Notice label changes: Early cans carried the brewery name. Then, probably because the Tivoli name turned many people off (Tivoli beer wasn't famous for its taste), the name was dropped. Finally, it was replaced with the premium designation—but to no avail. The brewery is now defunct. Tivoli spelled backward is "i lov it." But few did.

Busch Lager's short-lived introduction left collectors with a real tough can to find, but another of A-B's experiments with popular-priced beer produced some even tougher cans. Six different can designs were developed for a beer they called Busch Beer and these were sold only on a very limited basis in test markets to see which can design sold best. Because so few of each of these can designs were ever made, only a few collections contain them. Fewer still are those who can boast the complete set. Today, of course, the mammoth brewer's low-priced brand is Busch Bavarian. (Schlitz' counterpart is Old Milwaukee.)

A budget beer you won't find on the market anymore

(but you may find in dumps in the midwest) is old St. Louis. It was made by Griesedieck Bros. Brewing, but they tried to hide this fact from the public. On the can it said that it was made by Lami Brewing. A strange name? Not to those who know St. Louis. The Griesedieck brewery was *on* Lami Street.

Rheingold used the same ploy when they came out with Gablinger beer. It was supposed to be a product of Forrest Brewing. Yes, Rheingold was on Forrest Street in Brooklyn.

Although some brewers have tried to remain incognito, many others have made a Big Thing out of where their beer comes from in order to play up to civic pride.

There has never been a shortage of beers bearing brand names that feature cities and states—only a shortage of beerdrinkers willing to keep them all in business.

Billings. Boston Old Stock. Butte Lager. California Gold Label. Camden Lager. Chester Pilsener. Chief Oshkosh. Chippewa Pride. Cincinnati Burger Brau. Crystal Colorado. Dakota. Denver. Duquesne. Frankenmuth. Frisco. Great Falls Select. Kato (for Mankato, Minnesota). KC's Best. Kentucky. Manhattan. Menominee Champion. New York Special. Oconto. Old Milwaukee. Olde Virginia. Olympia. Phoenix. Potosi. R&H Staten Island Light. Reading. Regal New Orleans Famous Premium. Rhinelander. St. Louis Pilsener. Santa Fe Lager. Sheridan. Sun Valley. Tacoma. Tahoe. Tex. Texas Pride. Trenton. Utica Club. Virginia's Famous. West Virginia. And more. Many more.

But not all of these brands were actually made where you'd think. KC's Best, for example, came from Chicago and Omaha at various times. But never from K.C. Manhattan came from three different states—Connecticut, Pennsylvania, and Illinois—but never from The Big Apple.* The same went for Phoenix; it was never an

*Speaking of which, even Big Apple Beer didn't come out of New York. It was produced in Reading, Pennsylvania and Hammonton, N.J.

SHERIDAN can bearing those four magic words, "Internal Revenue Tax Paid" had to have been produced before March 1, 1950.

PRIMO, a Hawaiian brew, is the only canned beer the 50th state has ever produced. (Alaska has yet to come out with one.)

PROGRESS is the only brand of beer to come from the long-dry state of Oklahoma.

HAPPY HOPS was a recurring theme with the Grace Brothers Brewing Company. This is the can that started it all. Other Grace Brothers brands carried the Happy Hops symbol for years.

HALS BEER had the Old Masterpiece motif on the outside. Inside? Well, let's just say this brand is now obsolete.

HITT'S SANGERFEST gives you songs to sing while you're beering away the hours.

LAWMAKERS drink beer, too. And so does everybody else in the District of Columbia. Hence these brands out of the past.

Arizona brew at all—it came from Florida, Michigan, and New York. Los Angeles was the only home Santa Fe Lager ever had.

No wonder so few of these beers ever amounted to much. Matter of fact, of all the beers named for cities and states listed above, only Old Milwaukee and Olympia are powerful factors in today's beer industry.

Still, Old Milwaukee is only one of many brands that have carried the name of that beery city. Others include Milwaukee Brand, Milwaukee Premium, Milwaukee Valley, Milwaukee's Best, and (get this!) United Milwaukee Lager. Too bad only a couple are still around.

(The same goes for Wisconsin Beer, Wisconsin Club, Wisconsin Gold Label, Wisconsin Holiday, Wisconsin Lager, Wisconsin Premium and Wisconsin Private Club.)

P.O.N. and P.O.C. are two more brands that have played up to civic pride—literally—in their quest for sales. The initials stand for Pride of Newark and Pride of Cleveland.

While hopping around the country picking up on all the local brews (and their cans) one should always keep his eyes peeled for cans bearing a Happy Hops emblem. This insignia was carried by many of the many brands produced by the now defunct Grace Bros. of Santa Rosa, California. One of the most desirable of these is Happy Hops Beer, the can that got the whole Happy Hops phenomenon off the ground.

Though much more beer makes its home in 12-ounce cans than any other size, beer cans that hold, 7, 8, 10, 11, 14, 15, 16, 24, 32 and 128 ounces of the amber frothy are also to be found—and treasured.

Especially the 128 ouncers (gallons). Big beer cans are quite common overseas, but less than thirty U.S. brands have come out in these outsized containers. The hardest to find are Gettelman Bock, Koch's Deer Run Draft Ale, Blitz Weinhard, Grace Bros., and Brew 102. Only one or two samples of some of these enormous cans are known to survive.

Good beer and good times go together like, well, good beer and good times. Small wonder, then, that some brands have borne labels that make a big thing of partying. The beautiful full-color illustrations on Hals Beer is in honor of the famous Dutch artist, Frans Hals, who was famed for his paintings of people drinking and carrying on in 17th century taverns. Hitt's Sangerfest went a step further. The Hitt's can bears the words of barroom songs like "Drink to Me Only With Thine Eyes" and "Camptown Races." Sangerfest, by the way, means song festival. No doubt these cans inspired many a beery baritone. (Which may explain why Hitt's is no longer produced.)

9

The rarest cans of all—or the most common?

"Freak cans are not rare, not valuable, and many are just fakes which are being produced by the thousands to rip off the unsuspecting beer can collector."

Thomas Toepfer, Beer Can Collector
Aurora, Illinois

Experimental and short-run test market editions of beer cans are rare. There's no question about it. They rank right up there with the long-gone cans of the 1930s and 1940s in terms of desirability and all-around toughness.

So-called "mistake" cans are a different story. Take the can Bob Feldwisch has. It's a Budweiser overstamped with Dad's Diet Root Beer. And it's still full. (But with which?) The real problem is determining whether this can is the result of a bona fide manufacturing error or just some high jinks along the can assembly line.

More common is finding one can in a sixpack with the pull-tab on the wrong end. Other "mistake" cans you'll find floating around trading sessions have one or more (or all) of their colors missing (or shifted out of register), doubleprinted labels—or even the label on the *inside* of the can.

MISPRINT! Whether done on purpose or the result of a bona fide manufacturing error, each of these Budweiser cans rates a second look. Missing colors, upside-down cans, and multi-brand labels have never been particularly rare. And, now that they're being purposely manufactured, they're getting less so all the time.

SOMEHOW, each of these cans missed part of its paint job—but can you tell what brands they were supposed to be? The Stroh's is easy, but the other two aren't. The answer: Tivoli and Schmidt's of Philadelphia.

Here are some facts to remember about "mistakes":

1. Because of the high speed of their can making machines, few can companies meet their quality control standards on more than 95% of the cans they make. Which means that 5% of their production winds up in the scrap bins for recycling. Or for sale or trade to the unwary collector.

2. To be an authentic misprint, a can must have gotten past the inspectors, been filled with beer at the brewery and actually offered for sale on the retail market.

3. Cans with the label on the inside have usually been intentionally manufactured that way to be used as production markers on the can line to show where a certain count of cans ends.

4. Workers in can manufacturing plants have ample opportunities to produce just about any kind of "mistake" they care to.

This last point is the most distressing. One collector reports that he has been solicited by can company employees to buy "freak" cans in large quantities. "I can furnish all you want," the collector was told. "I'll make them to order."

Sad but true: A lot of collectors have been taken in by purposely-made freak cans. Some have paid dearly for them, too. They've traded 3 for 1, 5 for 1, even 8 for 1 to get them. One young collector was even seen trading four fine Fauerbach conetops for an upside-down Schmidt's can. Others, spotting them at flea markets, have even paid cash for what they thought were incredibly rare flukes. How many more cases of them do you suppose the flea market folks have back in the trunk of the car?

Hearing all this brew-ha-ha over cans that are probably phoney, one long-time collector says, "In my opinion, a misprint would never be worth as much as a regular can of the same type."

Gil Brennell offers the following explanation for the brand-on-brand cans you'll encounter. (Falstaff's label printed over Budweiser's. Or Peosi's over Miller's.) "If the sheet of tinplate that the first labels were printed on was not perfect, it was probably pulled before it went on to be slit and made into cans. Since the sheet was to be recycled anyway, it was used to check the printing setup for the second brand rather than using a fresh piece of tinplate. Later, this double-printed sheet could have been used to check the alignment of the slitting machine and then sent through the machinery that makes the can bodies to let the shipping department know where a certain count of a specific brand was ending—or perhaps to mark the end of a shift's production."

The moral of all this is obvious: Exercise extreme caution in trading for misprinted cans. That one-of-a-kind treasure may be one-of-many. Don't be one of many to fall for it.

10

Six pack, twelve pack, forty-one pack

"My favorite series of cans is the Rainier Jubilee series. As soon as my wife lets me move into a warehouse I'll try to collect them all."

Jeff Berg, Beer Can Collector
Rawlins, Wyoming

A schoolteacher up in the State of Washington undertook a monster task. He decided to figure out how many possible variations there are in the Rainier Jubilee series of cans that came out of the northwest in the fifties.

"Actually," says Bill Odell, "they came out in the thousands. You see, there were three sub-series within the series. The first one appeared in 1955. There were six basic can designs and they were made in eighteen colors. That's 108 cans right there. Of course, they were made both at the Seattle and Spokane breweries, so that makes 216 different cans. Each was produced with three word variations in the gold stripes, so that runs the number of combinations up to 648. But that's just the 12 ounce cans. There were another 324 combinations of 16 ouncers that came from the Seattle brewery. So that makes a total of 972 combinations in the first sub-series."

A QUINTET OF CANS from the beautiful Schmidt series of outdoor scenes. These cans are currently on the market, so any collector can get the entire set with a bit of judicious trading. (Or buy it, if he lives in Schmidt country.)

COMMEMORATIVE CANS come both in singles and in sets. The Schlitz can is just one of several the brewer put out to spotlight its brewery locations. (Longview is in Texas.) A-1 put out a special run of cans to honor its hometown's centennial. The Pittsburgh scene is just one of several you can find on the back of Iron City cans.

Odell—and Rainier's Jubilee can combinations—goes on. And on. The final tabulation reveals that there are a grand total of 11,052 different variations to be collected.

Nobody has them all, of course. Nobody even comes close. And, fortunately, most of the other can series that have come onto the market are more reasonable in their numbers.

The famous Schmidt outdoor life series, for example, has only a couple of dozen different scenes in it. Obtaining a complete set is well within the realm of possibility. All you have to do is hang around a liquor store up in Schmidt Country (the upper midwest) for a while. Or scour the roadsides. (Or come up with a couple of dozen traders needed by a collector who lives up there.)

THE GRETZ CAR SERIES includes a dozen cans, each with a picture of a different sports car on the back. The cans—like the cars pictured thereon—are out of the 1950's.

Another brewery that has a penchant for putting ou
sets of cans is Iron City. So far, they've had a series o
cans of Pittsburgh scenes, another of New Jersey scenes
still another of Christmas scenes—plus a vast array o
sets commemorating various aspects of the sportin
world. Usually, their emphasis is on Pennsylvania athlet
ic teams. In '74, for example, they put out nine differen
cans celebrating the Pittsburgh Pirates. One can carrie
the home schedule. Another bore the record of the Pi
rates' World Series play. It took two cans to bring us th
'74 Pirates team roster. (One for pitchers and catchers
the other for everybody else.) And so on. Iron City bega
putting out its many series of cans just a few years ago, s
most of them are still fairly easy to find. And more kee
appearing all the time. By now, Iron City has doubtles
learned that beer can collectors will gobble up 10,000 o
so of every different can they can make. So there's no en
in sight.

Drewry's of South Bend, Indiana, has also put out
number of series of cans, but that was a bit longer ago, s
they're a mite tougher to find. They had a sports series
an astrological series (six cans, each carrying two signs o
the horoscope), and a character series (telling how yo
can tell a person's personality by the shape of his face
nose, eyes, etc.).

Tougher yet is the Gretz can series. There are a doze
cans in this set, each depicting a different sports car
Unfortunately, getting your hands on them all is onl
slightly more difficult than finding the entire Rainie
Jubilee series.

The same goes for another Gretz series, the Toone
Schooner cans. Each can in this set from the 1950'
carried a song ("Good Old Summertime," "Rosi
O'Grady," etc.) and colorful stylized drawings of peopl
living it up in the Gay Nineties.

Living it up was also the theme of a series of Meiste
Brau cans from the late 1950's, but the subject wa
treated in a more contemporary manner. This "Happ

TOONER SCHOONER was another series of cans out of the 1950's. Made by Gretz, each can in the set carried the words to a song on the back.

PEOPLE HAVING FUN. That's the theme of this series of Meister Brau cans from the late 1950's.

Day" series, as it's known, features realistic full-color illustrations of people fishing, skiing, bowling, hamburgering, etc.

Another superset is the Parti-Quiz cans from Esslinger. There are believed to be at least ten varieties of these cans which were supposed to liven up your party with fabulous facts like "Snakes' teeth are called fangs," "The moon revolves on its own axis," and "Mississippi won the Cotton Bowl in 1956." One fact they don't tell: The venerable Philadelphia brewery went out of business in 1964.

11

Beer cans are where you find them

"You learn to read labels along the highway at 60 miles an hour—well, 55."

Rich La Susa, Beer Can Collector
Carol Stream, Illinois
(as quoted in *TIME.*)

You know the old joke about climbing to the top of a mountain that, supposedly, has never before been climbed—and finding an empty Coke can there? Not true. It would probably be Budweiser.

It's a fact: Beer cans *are* everywhere. Clever collectors know this and that's exactly where they look for old beer cans—everywhere.

The roadsides of America are a good place to start. The immense quantity of beer cans that are to be found along roadsides (and the thoughtlessness of those who tossed them there) appears to be unique to this country. A collector who walked the byways of England and Germany hoping to find some beauties for his collection came up virtually empty-handed. Matter of fact, he reports having encountered very little litter of any kind.

Well, here in America, we've got it. And beer can collectors are the only ones who both appreciate litter and help to remedy it by picking the roadsides clean.

Old lovers' lanes are another good place to check out for old beer cans. So are picnic grounds. When nobody's looking, just mosey on over to the trash can and start going through it. If a lot of the people using the picnic area are tourists, you've got a good chance of finding some brands they've brought with them for other areas.

Just wandering through the woods can be exceptionally productive if you're lucky enough to come upon an old picnic site that has been left pretty much undisturbed for a number of years.

But, as Jim McCoy says, "A can in the hand is worth two in the bush." So you've got to be sharp-eyed as you make your way through the underbrush. The best time of the year for woods wandering, of course, is early spring because of the lack of foliage that could hide those two cans in the bush from your sight.

The cans you'll discover in your outdoor searches will often be somewhat rusty and faded. Regardless of where they're found, these cans are known as "dumpers." And if it's a brand you don't have up on the wall, a dumper is a welcome addition.

For *real* dumpers—and to find old cans in large quantities—many collectors search out old dumps. Having spent a great deal of his adult life rummaging through the dumps of Washington state, Bill Mugrage has become known as the King of the Rusters. Premium Bill (as he calls himself) is famous for his standing offer: "I'll send you 25 pounds of Rainier rust for any can I don't have."

Another dump expert, Tim Zgonina of Radom, Illinois, offers these hints to dump diggers: "The best dumps are those that haven't been used for many years. Finding them isn't too hard if you know what to look for. Keep your eyes peeled for small patches of white color. All dumps have white things in them and it shows up well especially in the dead of winter. Bluffs along streams are good places to look, too. People often back their trucks up to these bluffs to dump their junk over the edge."

While out deer hunting in the woods of northern Wisconsin, Bob Bohmann came across a dump that made him forget all about Bambi.

"It was solid cans!" he says. "I came home with 53 cones and would have taken more if they hadn't been frozen in the pile." Subsequent visits to this Mother Lode have turned up around 3000 cans—and they weren't just Bud and Schlitz and Falstaff, either. "We found Menominee Champions, Silver Creams, Old Imperials—all conetops—plus some great flattops like Rahrs All Star, Oconto, Goebel, Edelweiss Cheery-Beery and a lot more."

After a little hiking and trash-kicking around an old resort, Earl Myers had a good haul for himself, too. "Those trash piles must date back to the 1930's," he says. "I brought home 39 different obsoletes including Braumeister and Berghoff conetops plus E&B Brew 103, Trophy, Yusay, and Edelweiss, just to name a few."

On a Canadian fishing trip with his father, Robert Meimer saved the cans his dad and some other men emptied. Then a pretty scary thing happened. "On the last day of our trip, the maid threw out all the cans I'd saved, so I asked where the dump was. It turned out to be in the woods, so we went to check it out. *Jackpot!* Not only did I get all my cans back, but I got many more!"

Those collectors who have logged a lot of hours in dumps say the best dumpers come from dumps that are shaded (reduces fading) and have good drainage (reduces rusting). When there are several layers of cans, the cans in the second layer from the bottom generally turn out to have survived in the best shape. The layers above protect them from the elements and the layer below takes the punishment from standing water. So be sure to dig deep.

One caution: The cans you find outdoors may be serving as homes for Small Creatures. Insects are the most common residents of old beer cans, but mice and snakes have also been known to inhabit them. As have entire families of hornets. Best idea: Thoroughly wash out your dumpers before you bring them inside your house.

Another thought: Equip yourself with a stick when you go dumping. After all, you don't want to go poking

DUMPING is a popular way to unearth rare old cans for your collection. Bob Stutzman always carries a stick when he pokes around in dumps, a recommended procedure.

around in that pile of broken glass, jagged metal and (maybe) angry snakes with your hands.

Two more good things to have along when you go dumping are a map (preferably an old one, to help you locate long-lost dump areas) and something in which to bring home your treasures. Many collectors use gunny sacks or plastic trash bags, but others point out that many dumpers are pretty delicate and suffer mightily when they rub against one another. Better to bring along a flat (empty beer case).

Yes, there *are* cans in them thar hills of trash. And they're well worth searching through. But after you've exhausted the dumps in your area, then what?

You play detective. You sit down and ponder these key questions: where have people been since 1935? Where have they gone to drink? Playing detective has really paid off for a lot of collectors.

Example: Tim Zgonina uses USGS maps to find old homesteads that are now uninhabited. The cans he's found inside and under these old houses (and the barns and sheds that are invariably nearby) are usually in good condition.

One shed he discovered produced a real prize—a quart Growler Manhattan Cone. Nearly every homestead had a nearby dump, of course. So that gives you two places to search—inside and out. Check out those old abandoned cars too, including under the seats and in the trunks.

Ohioan Tom Kirker does nice detective work, too. He searched through an old dance hall, discovered that the foundation had large holes in it, and crawled in. His reward: Old Bavarian conetops, some really old Pabst Export and Waldorf Lager flattops—and a lot more.

The hearts of all dedicated beer can sleuths beat a little faster whenever an old building is being torn down. If it was built since Krueger introduced the American construction worker to beer in cans, chances are good that it contains treasures. Seems that while they were putting those buildings up, a lot of carpenters were also slugging beers down—and stashing the empties right there inside the walls. The beautiful part of finding cans inside the walls of an old building that's being torn down is that, having been protected from both light and moisture, they're always in terrific condition.

While on vacation in North Dakota, Rollo Carlson learned that an old railroad station was being torn down nearby. Sure enough, when the walls were ripped out, hundreds of old beer cans came tumbling out. Carlson's haul: Thirty-three cases of flattops (most of which carried the Internal Revenue tax stamp) and 88 conetops!

Old railroad stations have provided similar bonanzas for other collectors, too. Bob Myers made one of the all-time great finds while rummaging around an old abandoned station in upstate New York. Seems that some long-forgotten stationmaster had killed time by drinking beer and tossing the empties through a hole in the ceiling. The attic provided Myers with cases of rare old Red Fox, Haberle, Tam O'Shanter, Fitzgerald and Beverwyck cans.

Back in the early 1940's, author Jack Kirkland wrote a play entitled "Suds in Your Eye." A comedy, it all takes place in a junk yard. For some productions of the play, sets consisting largely of beer cans were built. In a demonstration of first-rate detective work, Bob Myers has located and taken advantage of two of these sets. He found them languishing in theatrical warehouses. Needless to say, many of these cans—current when the sets were built—are now real toughies. They now languish in Myers' collection.

While you're prowling around the countryside in search of beer cans, don't overlook any building that might have once contained a workshop or tool shed. Many's the Farmer Brown who kept his nuts and bolts and axle grease in beer cans with the tops cut off. Even if the tools are long gone, a few fine cans may have been left behind.

Many city-bound collectors, unable to spend much

BEER CANS WITH A PAST. Collectors find cans serving as containers for nails, axle grease, oil, and who knows what else!

time poking around in rural areas (and unable to find beer can-laden theatrical sets), have found recycling centers a good source of collectibles. (Like they say, one man's trash is another man's treasure!)

And, needless to say, abandoned breweries have been known to have more than a few cans rolling around in them. Rare is the brewery that has gone out of business precisely as it filled and shipped the last empty can in the place. When they close their doors forever, there's usually a few jillion cans left behind neatly stacked on pallets. Which is why their doors rarely stay closed forever—collectors find their ways through them to get at the goodies. Some of the ploys that have been employed by can crazies eager to gain entrance into locked-up breweries have been truly inspired:

"I'm writing a paper for my architecture course at _______ University about this fine old building; mind if I take a look around?"

"I've been asked to appraise this structure for a prospective buyer, please let me in."

"My grandfather, ____________________, worked here back in 19____. He's asked if I would come and take a picture of the room he used to work in before they tear the building down. You wouldn't deny an_____-year old man, would you?"

By now, it's fair to estimate that nearly every out-of-business brewery has been completely stripped of its cans. Occasionally, however, somebody gets lucky.

There was the time John Ahrens, the John Paul Getty of collectors, made a pilgrimage to the old Esslinger Brewery in Philadelphia. He discovered that it had been operated as a warehouse for the last decade. He asked a warehouse worker if there were any old beer cans still around. (Doesn't hurt to try!)

"You're just a little late," Ahrens was told. "We had 10,000 unfilled beer cans here for years. But just recently, a turpentine company bought them up, filled them with turp and relabeled them. I have one sample here."

Ahrens peeled off the label and there was an old Esslinger can. Beautiful!

Talk about detective work! Mike Davis really put some to work in his search for a Soul can. Mike's best pal is on Los Angeles' SWAT team, you see. And he knew of Mike's ardent desire to give his collection some Soul. While checking out a bar in the Watts area, Mike's friend didn't find the suspect he was looking for, but he did spot two cans of Soul Malt Liquor sitting on the shelf. The bartender said they'd been there for about ten years and probably wouldn't be fit to drink, but he still had to have fifty cents apiece for them. Sold.

There's no question about it: One of the best places to find old beer cans is in the older bars of almost any city. Especially down in the cellars. That's where Chicago firefighter (and canfinder) Clayton Tichelar made his big find. Seems he was talking old beer cans in a tavern and the owner recalled that he had seen some years ago and they might still be down there in the basement. Tichelar asked if the owner would mind looking. The owner replied that he was getting too old to bother with such things. So a deal was made: Clayton would do a little cleaning up in return for any cans he found.

A few sweeps of the broom later, Clayton Tichelar was the owner of 120 obsolete cans including *fifteen* which weren't even on the BCCA obsolete list! A few of the brands should give you the general idea: Waldorf Samson Ale. Kroysen. English Lad Ale. Rosalie Pilsner (from Church Point, Louisiana).

Of course, taverns in many parts of the country serve only bottled beer. To buy canned beer, you have to go to a liquor store (or grocery store, in some areas). And you should, because you'll find many a goodie by checking as many liquor stores as you can. Not only will that keep you up-to-date with new currents as they come out, you can get lucky in a liquor store just like you can in a bar. Maybe, just maybe, there's an oldie lurking way back in the cooler. Or a case of something old and awful the owner has put away in the backroom because he feels it's too old to drink. (When this is the case, the owner is generally more than happy to find somebody to buy it who doesn't *care* what it tastes like.)

If you're a regular customer, a liquor store will probably do whatever he can to help you with your hobby. Like put away a sixpack of every new beer that comes along. Or shake down the distributor for a case of that odd-ball brand that just changed labels—or went off the market.

Naturally, whenever you travel you *will* take along an empty suitcase, won't you? Better make it *two* empty suitcases. Because there are a lot of beers out there and you're going to want to bring home a sixpack or two of each—one for your collection, some for your collector friends, and a bunch to use as traders.

Ken Hiestand, a long-distance truckdriver, knows what he's talking about when he warns: "You have to be prepared to encounter some pretty lousy-tasting beer. Some cans make great collectibles, but the stuff inside them is difficult to drink." A feeling echoed by a young collector's father who laments that "I've drunk every raunchy beer in seven states for my son. But I'm willing to make the sacrifice." Now *that's* love.

And when you're not out there snooping through all the liquor stores from coast-to-coast, make sure your friends are. Ohio collector Paul Ladefoged says, "I ask all my friends to bring me cans from wherever they go whenever they take a trip. And they do. Everybody likes to help the criminally insane."

You can't hide your insanity from your friends, anyway, so you might as well let everybody know you're a collector of beer cans. Get the word out! Talk it up! You'll be surprised how many cans this will bring you. Often, it'll even bring cases of oldies that people stashed in attics, years ago, for no particular reason.

Newspaper coverage of you and your collection won't hurt, either. And chances are good that somebody across town will call to offer you some real beauties. "Just come on over and pick'em up anytime," they'll say. "They're just collecting dust here."

Who knows? Maybe, with a little persistence and luck, you'll even beat Minnesota collector Bill Henderson.

He found an ancient Gettelman can in the cargo compartment of a World War I fighter plane at an air show.

12

The beauty treatment

"You always know what the before is, but you never know for sure what the after's gonna be. . ."

Jim McCoy, Beer Can Collector
Denver, Colorado

A lot of beer cans you'll run into will be rusty graduates of the College of Hard Knocks.

That's why so many collectors have spent so many hours trying to find ways to restore unsightly cans to their original loveliness. A lot of cans have paid The Ultimate Price to further the cause, but all this tinkering has really paid off.

There are five kinds of restoration work that collectors employ in attempting to bring cans back from the critical list. Some are still in the experimental stage, a couple may or may not be fraught with danger to the health of the collectors who employs them, and a few are pretty sure bets to work.

They fall into seven general categories:

1. Soap & Water
2. Scouring & Scrubbing
3. Body and Fender Work
4. Chemical Potions
5. The Earl Scheib Treatment
6. Preventive Care & Maintenance
7. Weird Stuff

First, the good old soap and water treatment. As John Butera says, "When you find a can that is dirty or rusty, don't jump to any conclusions before you wash it thoroughly in soap and water. Often that 'rust' turns out to be heavily caked dirt." All the dirt, bugs, and sticks that are hiding inside may make a mess in the sink, but if you give your cans a good going over with a toothbrush and mild soap, you'll often be surprised at the results.

Some collectors soak their cans; but be careful. If left in water too long, the paint may soften and begin to peel.

Then there's scouring and scrubbing, a very dangerous area to work in (not for you, for the can), but one which many collectors have found useful. John Butera deals with light surface coats of rust with zealous scrubbing with 409 or Fantastik, making sure the cleanser he uses doesn't contain ammonia or acid. If this fails, Butera treads where others fear to—he scrubs lightly with steel wool.

If rust hasn't eaten very deeply into the metal, it can often be removed with a carefully-wielded pencil eraser. (But don't use an ink eraser; it's too gritty.) A few brave souls have also tried the Brillo/kerosene treatment. But you have to have a mighty light hand or the paint will suffer.

Emory paper can do away with the rust around the tops and bottoms of cans, and some recommend that you use masking tape to protect the label from accidental scratching. Dick Roser, however, says he's had masking tape pull some of the paint off a couple of his older cans and suggests you restrict the tape bit to fairly new cans. To really shine up those rims, of course, put a wire wheel on your grinder and go at it.

Now that we're getting physical with our cans, let's get *really* physical and muscle the bent ones back into shape.

But first, consider the sad case of Michael Jordan. He's just flushed a Shell's Pilsener can out from under a tropical shrub in Miami's Flamingo Park and was carying it home when it popped out of the shopping bag he'd stowed it in and threw itself under the wheels of an oncoming bus. Shaken, he scooped it up, took it home, cut off the bottom, worked a Dr. Pepper bottle into its badly battered body and rolled it back into shape. Sort of.

When we talk about the physical refurbishing of cans, this is what we mean. Rolling. Pressing. And blasting. And, since showoff beer drinkers so often crush their empties, collectors have devised a lot of beer can unbending procedures designed to force cans back to a semblance of their natural shape.

One method (which has met with only limited success) is to fill the crumpled can with water and freeze it. The water will push out the sides of the can while it expands. Sometimes not enough, though. And sometimes too much. Best to check the freezer frequently.

Another popular method of restoration involves the use of firecrackers. Safety factors and state laws make this a risky venture—so let's not go into it here. (If you'd like to go into it, be sure to read the hints—and warnings—in *The Beer Can* (Matteson, Illinois: Great Lakes Living Press, 1976). This BCCA-authored book also goes into oxalic acid restoration, but not without gloves.)

Air pressure—other than from fireworks—has also been successfully employed by some of the more technically-minded collectors on cans suffering from The Bends. For cones, you need a complete valve stem assembly including the round rubber-ball thingie approximately 1 inch in diameter. This will just cover the top of the conetop. Wrap several layers of cloth around the can (to protect from breakage), press the ball down on the top and give it one quick shot of air from a gas station airpump. If it doesn't work the first time, it probably won't work at all.

Gil Brennell, a heating and air conditioning guy by trade, is an old hand at pounding tin into shape. Which explains why he has devised a complex method for forcing flattop cans into shape with up to 125 psi of air. To accomplish this, you need all kinds of metal parts, thread rod, wingnuts and rubber gaskets. Gil is the only man in the world who can understand his method, much less put it to work, so let's just forget it.

When it comes to reconstructing beer cans, few can hold a candle to Tom Williams of Springfield, Illinois. And he has some pretty strong ideas on the subject: "Let's put the kibosh on firecrackers, soda bottles, freezing and all other barbaric means to de-dent cans."

Tom is an engineering technician, and his sophisticated method for repairing dented cans is diagrammed on page 127.

Now that all your cans are once again upright citizens, you may want to turn your attention to the removal of those rust spots that didn't yield to washing, scouring, or erasing.

What you need, obviously, is some kind of chemical that eats rust for lunch—also breakfast and dinner. Fortunately, there are two on the market. Each has both advantages and disadvantages.

First, Naval Jelly. It dissolves rust quite quickly. Most hardware stores sell it. (There is also an Aluminum Jelly which cleans and brightens aluminum.) When using strong chemicals like the members of the Jelly family, be sure to follow the directions. Otherwise, you may suffer burns on your hands. Worse yet, you may remove the paint on your cans. So don't use too much or leave it on too long.

Normally, a five-minute application of Naval Jelly will do away with a good deal of rust. Always start with a dry can, and always rinse it when you're through.

Metallic and red paints are particularly apt to react poorly to Naval Jelly, so be extra careful with these. If you notice running, fading, or peeling, rinse immediately.

Naval Jelly tarnishes aluminum, so be careful not to apply it to aluminum cans—or the aluminum tops some steel cans have.

As with all can restoration methods, be sure to try 'em

1. Grind at edge of rim. (But don't grind too much because the can metal under the rim will be needed later.)

2. Grind through rim at can joint. After grinding, end and rim will come away as shown.

3. Put can over 2" pipe held in vice and tap out dents with hammer. Tap lightly. Or use rolling pin. Or use both.

4. Cut a good end from another can, insert small screwdriver between rim and can and tap with hammer to open a space all around.

5. Crimp with needlenose pliers or crimping tool.

6. Insert new end into repaired can, allowing extra can metal left in first step to slide into space made in Step 4.

7. Tap new end to grip onto can. There. Isn't that better?

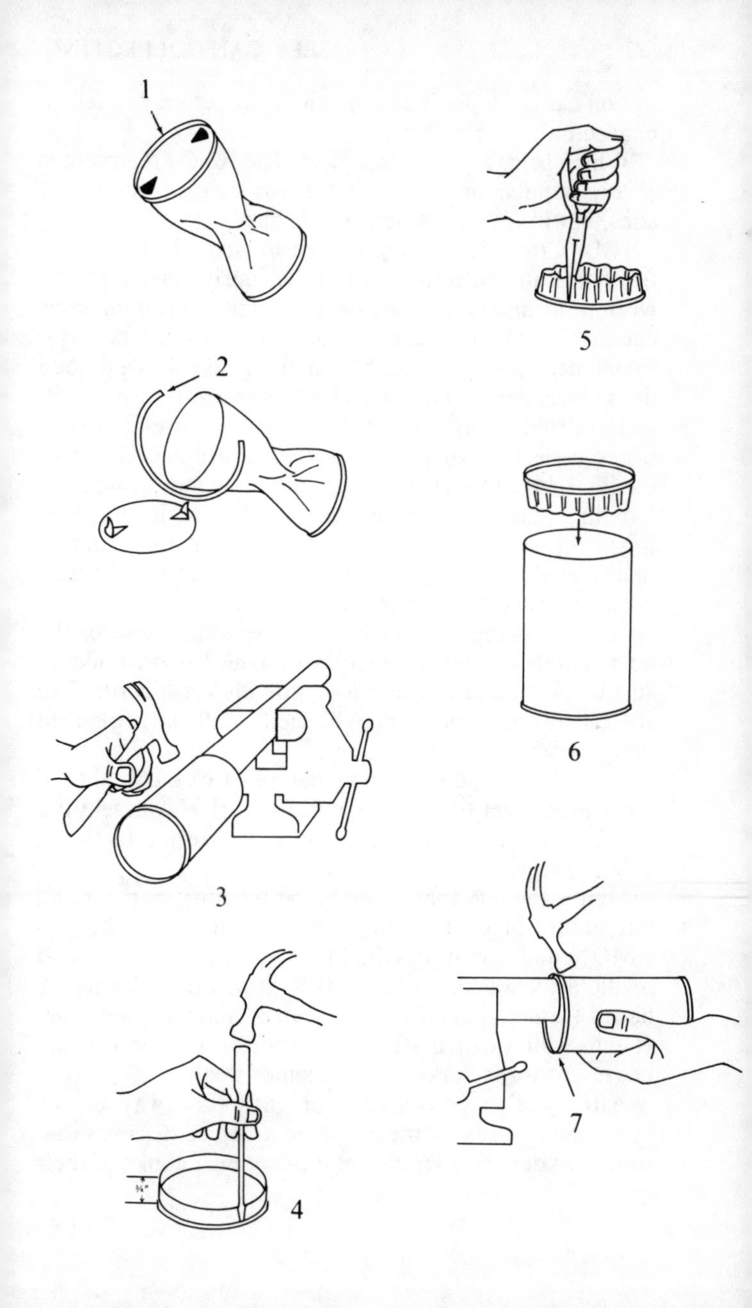
1
2
3
4
¾"
5
6
7

out on cans you don't care much about before you take a chance on your toughies.

Which brings us to oxalic acid. The Acid Treatment is a very popular method of doing away with rust on beer cans, but it has its detractors. Serious ones.

H.L. Cover is a chemistry professor when he's not being a can collector, and he seriously questions the wisdom of allowing youngsters to handle this dangerous chemical. "Oxalic acid is a violent poison," he says, "And neutralizing it with something like baking soda doesn't render it harmless. The neutralization product, oxalate ion, is just as lethal as the acid, itself." Cover urges collectors to recognize the hazardous nature of oxalic acid and suggests that only someone trained in handling dangerous chemicals should use it. Furthermore, he says, the commercial products that contain oxalic acid should be used only as directed for the purposes intended by their manufacturers.

After cleaning thousands of old beer cans with oxalic acid, another collector reports that he had to undergo surgery for the removal of a cyst on his vocal cords. The specialist who treated him blamed it all on prolonged breathing of oxalic acid fumes.

Considering the dangerous nature of oxalic acid (and what many feel to be the inadequacy of Naval Jelly), a number of collectors have been experimenting to find a safe substitute.

Here's the one that seems to be the most popular: To one gallon of water, add a 16 ounce bottle of REAL-EMON and a cup of salt. Mix well. Use as you would oxalic acid, washing your cans both before and after. A bonus feature provided by the water-lemon-salt solution: It tells you when it's losing its ability to remove rust because it turns darker as it becomes weaker.

After you've de-dented your cans (one way or the other) and de-rusted them, you're going to discover that some of your little jewels are missing big chunks of their paint.

To repaint or not to repaint? Whether or not to give a can The Earl Scheib Treatment *is* a knotty question. Some say it's downright immoral to pass off a repainted beer can as being anything but a counterfeit—whether you put it up for trade or up on your display shelf.

Others disagree. (Like Dick Adamowicz, known throughout the brewerians' world as Mr. Repaint.) And they go right on giving their old dumpers The Ultimate Restoration—a new paint job.

One thing's for sure: If you *are* going to repaint, you'd better do an awfully good job of it!

Here are a few tips from another repaint wizard, Bill Christensen, that will keep you from having to toss botched-up repaint jobs back into the dump:

1. Don't try to do the job with cheap brushes. Get yourself some red sable artists' brushes—a number 2 for the big areas, and the 00 and 000 sizes for finer work. A good investment.

2. Use good paints—and use them sparingly. Testor's is good for shinier cans; flat lacquers like Imrie-Risley Military Colors are just right for older cans.

3. Be careful to correctly match colors. Especially when you're working with white. Most white paints are whiter than beer can whites.

4. Matching shininess is important, too. Maybe even more so than matching colors if you don't want people to find out you've, uh, *enhanced* the can. If the paint dries too shiny, rub it with your thumb until it is sufficiently dull. If it dries too dull, apply a clear semi-gloss.

5. If you goof up, don't try to rub it off. Let it dry, then pick it off with a pin.

There. Now that you've engaged in a bit of antique restoration (as some prefer to call it), are you going to foist it off on the world as The Genuine Article? Better not try. Instead of being known as a beer can artist, you might wind up with a rep as a *con* artist.

Christensen signs his work on the bottom of each of his repainted cans and never tries to trade 'em as originals. He admits that many collectors prefer cans in their natural state to restored ones, but notes that "such people also generally like organic vegetables, the rhythm method, and planters' warts."

Once they've picked up a new can for their collections, most collectors just put it up on the shelf. But some apply various potions hoping to ward off further aging of their little beauties. Various brands of furniture polish have been utilized for this purpose, Johnson's Pledge, for example. So have auto polishes, sewing machine oil, and baby oil. And collectors have sprayed their cans with everything from clear fingernail polish to clear vinyl. Others scoff at all these practices, warning that over a period of years, yellowing could begin to set in.

Most agree that a can sitting on a shelf, protected from direct sunlight and excess humidity, will retain its original colors indefinitely.

And Joe Veselsky is probably right when he says that the only thing you should put on your cans is a feather-duster.

Now for the weird stuff.

If you just *have* to have bottom-opened cans (or what looks like bottom-opened cans) on your shelf, you can glue pull-tabs back on all those that lack same. Elmer's Epoxy Clear Waterproof Glue will do the job nicely. It dries clear, and the excess can be removed with denatured alcohol after it has set up.

If you have a conetopless conetop, you can give it a new cone quite easily. With tin snips, cut the top rim off your old can and file smooth. Cut an automotive additive can (STP, for example) in half, about 1 inch below the conetop, with a hacksaw and file off the jagged edges. Crimp the 1 inch area inward a tad. Then press them together.

Once again, a question of ethics is involved. Best not to mis-represent this type of restoration, either.

And best not to even *attempt* to freshen up rusty old cans by painting the tops and rims with shiny new metallic paint like some have. The result is always the same. The cans look just like little old ladies in tennis shoes.

13

1000 cans of beer on the wall

"My wife suggests this method for saving space when you display your beer cans. First, she says, cut both ends out of them with a can opener and set them on the floor. Then you jump on them just as hard as you can. Our divorce becomes final in ten days."

Ron Biberthaler, Beer Can Collector
Federal Way, Washington

There's beer in beer cans, you know.

This fact often is overlooked in the quest for more and more cans to add to your collection. Nevertheless, a collector frequently encounters beer while engaged in this hobby and most elect to remove the contents before putting the can up on the shelves. Long-time collector Jerry Weishaar has pondered this weighty matter at length and has what he believes to be a workable solution: "I drink it," he says.

This plan has several striking advantages. It eliminates the possibility of suffering the heartbreak of leakage. It also reduces the possibility of breaking your bank account trying to buy strong enough shelving to hold up a wall of full cans.

OPEN FROM THE BOTTOM! That's the cry of the beer can collector who likes his cans on display to look like they're full.

Another reason some collectors give for displaying only empty cans is that when you drop a full one it is likely to dent. Empty ones will just bounce.

You should always open your beer cans on the bottom, of course. Because they look better up there on the shelf. (Kind of like uncirculated coins.) Besides, top-opened cans collect dust inside them, say some zealots. (But who would ever know?)

For a conventional steel can, bottom-opening is no problem. A simple application of the old churchkey and you're in business. When Jim McCoy empties a can to set up on the shelf, though, he doesn't use a churchkey—not even on steel cans. He uses an ice pick. To prove that it was a full can when he got it.

Of course, there's no lip on the bottom of the new extruded steel and aluminum cans. Because there's nothing to hook your churchkey onto, you'd be hard pressed to open one with a churchkey. But it *can* be done. You can lay a churchkey across the bottom with the point where you want to make a hole, then, with all your might, force the point through the metal. (To work, the churchkey *does* have to be hard-pressed!)

George Eckert has come up with an easier way. "Get a wooden doorknob," he says, "and a 3/16 inch spike, nail or punch that can be screwed, soldered or brazed onto the brass end of the doorknob. After you've screwed, soldered or brazed them together, you've got a nice beer can punch."

Now that they're all empty, you've got to determine how you're going to display your cans. If you've got a lot of bookshelves, but hate books, you're in luck. But you'll have to have a *lot* of bookshelves—about 23 running feet for every 100 cans.

Chances are, you're going to have to become a carpenter for a bit. And it probably won't be the last time. As Illinois collector Dick Armstrong says, "I am forever building new shelves."

Now, what kind to build? It's an important decision to

A GOOD SHELF SYSTEM is attractive and makes efficient use of the wall space. This one is made of channels, brackets, and redwood lawn edging.

make because once you start you've pretty much set the style of shelf you're going to have to live with—and look at—for the rest of your beer can collecting life.

If you decide to go the all-lumber route, avoid using thick boards for the shelving, proper. When you're running your shelves ten or twelve rows high, that extra thickness can really eat up the space.

Of course, if you pick boards that are too thin, your display will have the old ocean wave effect.

One of the best shelving systems—maybe *the* best—uses metal standards (channels) and brackets. You simply mount the standards vertically up the wall every two to three feet (depending on the shelving you choose), then snap in the rows of brackets and lay in your shelving. It's easy. It's fast, And it's moderately expensive. But it's very sturdy and makes economical use of the space. You can use any kind of wood you like for the shelving, but a number of collectors like the looks (and thinness) of redwood lawn edging. Check into it.

If you decide to use penetrating walnut stain on your shelves, take heed. Several collectors discovered the hard way that it gives off fumes (or something) that, over a period of time, discolored the white paint on their cans. If you're going the stain route, be sure to put a sealing coat over it.

To make efficient use of your wall space, display your 16-ounce cans separately from the 12s and under. Other than that, you can organize your cans whatever way you like. Most collectors do the obvious: They put their cans up in alphabetical order, the domestics separate from the foreign. But not Paul Ladefoged; he goes by state and country—alphabetical within each, of course. Max Robb alphabetizes by *brewery* rather than by brand. Dave Nolan does none of the above. Concerned that his collection was overrunning his house, he stores his entire collection in his garage, leaving just 100 cans demurely on display atop his mantel. "They're my 100 favorites as of this moment. I keep changing them, of course."

Elmer Mick took just the opposite tack. "After seeing many collections, I noticed that the average display on shelves is about 10 to 13 cans high, so I decided to have The World's Highest Display of Beer Cans. I started building my display on the basement floor and ran it all the way up the stairs to the ceiling of the first floor. It's *forty cans high!*"

While St. Louis collector Mick goes up, Austin's John Zembo goes out. When Zembo ran out of room in his house to display his cans he did what any good Texan would do. He built a 425 square foot addition. Just for his beer cans.

Steve Baumgartner's display became too much to handle where he lived, too. So he moved. "We lived in a two-bedroom apartment when our baby was born," he says. "My wife refused to put the baby in the beer can room and I refused to move my beer cans. So we had to buy a three-bedroom house. One room for us, one for the baby, and one for the beer cans."

14

A six pack of beer can collectors —and their collections

"Don't worry if your collection is small and conetops are few. Remember, the mighty oak was once a nut like you."

Banquet Placemat, BCCA Canvention V,
Des Moines, September 13, 1975

A mortician.
An unemployed salesman.
A sixth grader.
A rare coin dealer.
A computer systems and programming manager.
A law book salesman.
Six collectors of beer cans. All very different. And, for each of them, there are 2,500 *other* beer can fanatics out there. They're all very different, too, except for one thing: a strange fondness for certain canisters of tinplate and aluminum.

Armin D. "Shorty" Hotz was one of the first to get the bug. When conetops originally came out, Shorty was immediately struck by the unique containers for beer. "I decided right then that beer cans were beautiful and I started saving them."

As the years passed, Shorty's friends kept bringing him more and more cans, usually conetops—and usually full. Shorty kept them all.

Once Shorty and his wife, Esther, discussed tossing out the whole bunch while they were cleaning up after one that had sprung a leak. Fortunately, they decided to give the cans another chance.

"But I never really took the beer can thing seriously," says Shorty, "until one day Esther put a clipping from the *St. Louis Globe-Democrat* on my desk. There was a picture of Denver Wright, Jr. and the St. Louis boys and an article asking if there were any other crazy people out there."

Curiously enough, the Hotz home furnishings store and funeral home in St. Peter, Illinois, had been buying advertising materials from Wright for 35 years. So Shorty called up his longtime business associate, asked him what it was all about, liked what he heard—and signed up.

It's one of the best things I ever did," says Shorty. "It opened up a whole new world for Esther and me. I've never been an aggressive trader. My main interest is the friendships that I've made through meeting other collectors."

Instead of making a big deal out of trading, Shorty just gives his extra cans to collector friends who need them. Invariably, he is remembered by these friends when they have extra cans.

Even so, Shorty's collection now numbers around 1000 cans. And Shorty, himself, is regarded as one of the most popular members in the BCAA. When "the St Louis boys" formed the Gateway Chapter, their bylaws limited membership to those people living within 100 miles of St. Louis, "*plus* Shorty Hotz."

SHORTY HOTZ. One of the world's first collectors of beer cans, Shorty gives Joe Strullmyer a few hints on the hobby. 17 year old Joe now has well over 1,000 cans. *Photo Credit: Dona Strullmyer.*

Budweiser
ON TAP

KENNY JERUE. He traded his first collection for a Cadillac and is hard at work on his second. *Photo Credit: Anaheim Bulletin.*

Kenny Jerue has long been interested in the contents of beer cans, but it wasn't until he got a job in the Maier Brewery in Los Angeles that the cans, themselves, started to fascinate him.

"I couldn't believe how many different cans Maier put the very same beer into. So I started taking home one of each just for fun."

Pretty soon, the inevitable happened. Jerue started adding non-Maier cans to his collection. And when he learned there were other collectors out there, he began trading by mail, sometimes spending up to thirty hours a week engaged in correspondence—and in packing cans for shipment.

All this effort really paid off. Jerue's collection grew to immense size and it continued to grow until A Strange Thing happened. Jerue fell in love with a Cadillac. Boy, did Kenny love that car! He loved it more than anything. More, even, than his beer can collection.

And so, suffering from a clear case of temporary insanity, Jerue *traded his entire collection for a car.**

But fate didn't let Jerue stay out of the beer can collecting world for long. One day he returned from a spin in what had once been his beer can collection and discovered a package. It was some cans he'd initiated a trade for before he'd decided to give up collecting.

Before he knew it, Jerue was hooked again. "Luckily," he says, "the guy who bought my collection so I could get the car didn't take my traders, so I had some raw material to work with." Furiously, Jerue wrote, firing off trading letters by the dozens. "I've been a busy boy," he says. "Besides, when a lot of my old collector friends found out I was back on the collecting scene they sent me cans to help me get started again."

---

*Actually, the guy with the car didn't want to trade it for beer cans, so Jerue had to use a middle man. He found a collector who was willing to trade money for the cans, then he traded that money for the car.

EDDIE MACHACEK. This 12 year old has not only put 2,000 cans up on the wall—he's got his parents collecting cans now, too! *Photo Credit: Rochester Post-Bulletin.*

Yes, Kenny Jerue is definitely back on the collecting scene. "I've even traded for some of the very same cans I had in my original collection!" he says.

Jerue's collection is back, too. "I have 5,500 now. That's more than twice the size of my first collection." And where does he keep his collection? "I keep my cans here in the garage," Jerue smiles. "Instead of the car."

Here, in Jerue's opinion, are the five toughest cans of all: Cremo (by Vernon Brewing), DeSoto (by Tennessee Brewing), Triple Crown (by Ballantine), Fredericks Four Crown (by South Side Brewing) and G.E.S. (by Drewry's).

Since his stint with Maier, Jerue has operated a business selling novelty watches and been a salesman for a large manufacturer of industrial V-belts. Currently unemployed, Jerue is a fulltime beer can collector—and emptier.

Schoolwork keeps sixgrader Eddie Machacek (Rochester, Minnesota) from being able to dedicate himself to fulltime beer can collecting, but he has helpers.

"I started by saving those Schmidt scenic cans," says Eddie. "Then I branched out and started collecting all the other cans I could find. When I got up to about 470, my Mom and Dad got involved."

At the Machacek house, it's been a lot like the junior-gets-the-electric-train-but-Dad-plays-with-it story. "My Dad joined the BCCA in his name," says Eddie, "because he thought I was too young."

With all three Machaceks pitching in, Eddie's 470 can collection has now gone over the 2,000 mark and includes 140 different conetops.

"Dad and I go dumping, Mom straightens 'em out, and Dad builds shelves to put 'em on," twelve year old Eddie reports.

Helped considerably by the discovery of 7 or 8 cases of Fountain Brew in an old dump, the Machacek collection has pretty much overrun its room in the basement. "Pretty soon we're going to have to thin 'em out," Eddie says. "We'll probably pull the aluminum look-alikes off the shelves and put them into storage."

Space limitations are creeping up on Bill Christensen, too. The Madison, N.J. collector glances over his cans. "I've got about 4,500 of them now, including 300 conetops and 50 quarts. As I run out of space I find myself more and more interested in New Jersey cans, particularly Ballantine."

Christensen now has 142 different Ballantine cans, but even so he refused to take beer can collecting too seriously. "I am constantly amazed by the growth of the hobby whose principal appeal I still believe is that it is so utterly ludicrous!"

A rare coin dealer by trade, Christensen provides a good insight into beer can collecting and how it's developing: "Five years ago, there were just two or three hundred collectors. So if there were only a couple of cases of a particular can, it was considered to be a moderately scarce can. Now that there are so many more collectors, any can available in that quantity would have to be regarded as being very, very rare."

But there's hope. "There are still plenty of mint oldies around somewhere," Christensen assures us. "Old collections keep appearing in the damndest places, plus there are lots of distributors' leftovers, display cans, and just plain cans lurking everywhere. It will take a while to find them, but they will show up—and will continue to show up for many years to come."

Why is a rare coin dealer involved in this beer can thing? "Since my profession is a hobby for most folks, I

BILL CHRISTENSEN. He has 142 different Ballantine cans in his collection. But thinks the idea of collecting empty beer cans is "ludicrous."

must look further afield for my own hobby. That's why I collect toy soldiers. Some call them Military Miniatures because they're embarrassed to admit they collect toy soldiers. I am not embarrassed. Nor am I embarrassed to be a collector of beer cans."

Besides a 143rd Ballantine, the cans Christensen would most like to get his hands on have not, as yet, been produced. "I'd like a Hopping Toad Malt Lager can—particularly one from the Famous Swamps commemorative set. Speaking of commemoratives, why isn't there one for the 910th Anniversary of the Norman Conquest? Or the 1500th Anniversary of the Fall of Rome? Most of all, though, I'd like one of those Polish gallon conetops. The ones with their spouts on the side."

If there were a Polish gallon, Denver's Ron Moermond would probably have it in his collection.

After all, he has well over a hundred foreign gallons (or their equivalent) plus what may well be America's largest collection of U.S. gallons.

Why gallons? "I like big cans," he says. "They've got some bulk to them."

Ron also likes 12 ounce U.S. cans and all sizes of foreign cans. At last count, he had 3500 U.S. cans (340 of which are cones!) and 1100 foreigners living at his house.

It takes up two rooms in his basement, but Ron is very happy with his 4600 can collection.

"It sure beats coin collecting," he says. "You can't keep a good coin collection in your home; you've got to visit it at the bank. Beer cans are different."

For Ron, it all started in early '71 when he lived with a bunch of bachelors. "They all drank a different brand of beer, so I saved one of each, just for the fun of it." From there, he branched out just a bit. "I began to wonder how

RON MOERMOND. With his fabulous collection of U.S. gallons plus one hand-painted fake: Iron City.

GENUINE
BEER
DRAFT
BEER

many different Colorado cans there were. Pretty soon, I started wondering how many different cans from *anywhere* there were, so I started saving those, too."

His work with computers involves a fair amount of travel and this has helped Ron's collection considerably. "I always take along some traders and get in touch with collectors everywhere I go."

"But the thing that gave my collection the biggest boost," says Ron, "was the time my Mom, through her card club, found out about an old collector up in Iowa who'd lost interest in beer cans long ago and quit the hobby. I went up and found the man. Pretty soon, I had myself 500 additions to my collection—all different, of course—plus 400 tough traders, all from before 1960."

"Yeah, I buy cans," says Ron. "I know a lot of collectors don't—and won't. But I figure that if a can is fairly priced and I can afford it, why not? But I've never sold a can. And I *won't.*"

Mostly, though, Ron gets his cans via the popular trading route. And from co-workers who travel overseas. "I get a big kick out of their reactions to my collection," he says. "They laugh at you. They say you're nuts. But when they see them, they'll spend hours studying them—then ask if they can bring over a friend. And when they see a strange new can in their travels, they always remember."

As for the future of the hobby, Moermond sees nothing but good things ahead. Plus one big black cloud.

"The collectors may wind up saving the small breweries that are left. There are so many of us now that a small brewery can move its product just by changing its can design. And they're starting to realize this, too. Just look at all the new cans that are coming out. I think we're stimulating a helluva lot of breweries. Just look at these Ortlieb's cans—they say 'Collector's Series' right on the front of the cans!"

Further proof of the growing influence of the hobby, Ron feels, is the fact that the president of Acme Brewing came to the Canvention in Des Moines to receive the BCCA's Can of the Year award. "It's gotta mean something when a guy like that flies out just to pick up a $5 trophy—and then spends three days drinking with us!"

The black cloud, Ron feels, is the questionable future of the beer can, itself. "If the anti-litter people force beer cans from the American scene, we'll still be around, of course. And we'll still find old cans here and there. But it won't be the same."

According to Moermond, six of the toughest cans in his collection are Easy Life, Connecticut Yankee, Bay State Ale, Roger Wilco 199, Krueger Bock, and Hornung. The toughest sixpack in the world, in his opinion, would be Budweiser Bock, Ambrosia, White Bear, Ronz, Connecticut Yankee Ale, and a Grace gallon.

Notice that he doesn't feel he has any of the world's six toughest cans in his collection. That's the way it is with all collectors. It's the green grass on the other side of the fence syndrome. If you've got it—no matter how hard it was to get—a can can't be as hard to get as a can you haven't yet been able to get.

John Ahrens has been able to get more of them than anybody. But he's not sure exactly how many he has. It's easy to see why: "I've got over 7,500 'regular' U.S. cans, over 400 conetops, over 2000 foreigns, and nearly 200 gallons, both foreign and domestic."

That's over *10,000* beer cans. And Ahrens isn't about to spend all the time it would take to find out exactly how many. "I'll count again next time I move," he says.

The law book salesman from Moorestown, N.J., got into collecting cans like so many collectors do—by accident. His senior year at Yale, Ahrens and several of his beerdrinking buddies decided to see how many different

JOHN AHRENS. The John Paul Getty of the beer can collectors. He has over 10,000 different cans. *Photo Credit: Sam Nicolla, Philadelphia Bulletin.*

brands of beer they could find. Mainly just to try out a lot of different brews. But also to help decorate their dormitory rooms. (His dorm had these nifty little ledges high up the walls about the height of a beer can from the ceiling, you see.) Then, when summer vacation came, Ahrens was somehow elected to de-decorate his room. Instead of throwing all these cans away, he kept one of each that was different. And he doesn't quite know why.

A decade later, John (and wife and kids) shares his suburban home with over 400 cases of empty beer cans. they're everywhere. In the den. In the garage. In the attic. "Someday," John says, "I hope to museum it."

With so many cans, you'd think Ahrens has put no restrictions on what he considers to be a different can. Not so. "Basically, I'll only keep a can if the can part is different. I do not save different tops—flattop vs. pull-tab, etc." he says. "Nor do I save different can companies, though once I did. And I do not save tax stamps. No, I do *not* save everything."

Try telling that to his wife, Brenda.

15

For kicks, for bucks—or for both?

"A beer can collector will trade with anybody—but he won't sell his Soul."

Charles Kuralt, CBS Evening News
March 25, 1975

Denver Wright, Jr. has had his home burglarized. So has Ted Stolinski. In both cases, they had a lot of valuables ripped off. But no beer cans.

That's because beer cans are worthless. Or are they? As the hobby grows, the point is being debated more and more.

Back in 1971, Ernie Oest, then a big beer can tycoon, said: "To many, this type of collecting is stupid. Stamps and coins can be bought and sold, but cans have no value. Besides, a true collector doesn't count what he can get out of it moneywise—just what pleasure it gives."

Four years later, Oest sold his collection.

Nevertheless, many collectors would not—under any circumstances—buy or sell a can or give them monetary values. "The best things in life are free," says one collector, "and for years beer cans have enjoyed that status." Charlie Miller feels the same way: "It takes away the fun if you buy a can."

Tom Teague says: "I readily admit to buying a significant portion of my collection and traders—a sixpack at a time. But to assign dollar values to my cans and the experiences they represent? Not on your tab top. If you do, you miss a big part of what collecting is all about. . .just plain good times."

Buying sixpacks doesn't count, of course. What's at issue here is whether or not buying and selling of cans for more than their retail value should be engaged in.

It is true: Most collectors *don't* sell cans. As one says, "Why would I want to sell a beer can? I'd lose a trader. All I'd get for it is money. I don't collect money. I collect *beer cans!*"

But couldn't he use that money to buy beer cans? Sure, but not from a collector. The people with beer cans for sale aren't collectors—they're *dealers*. And, because they're in it strictly for bucks, you'll be hard-pressed to find any deals when you do business with a dealer.

Example: One California outfit will sell you your choice of 25 brands for a buck each (minimum order: $6 worth). The flyer they send out doesn't say so, but all are current cans. You could buy any one of them in a supermarket right now at the usual sixpack prices. *Full.*

Another guy is willing (tickled to death would probably be more accurate) to send you sixpacks of empty King Snedley's, Buffalo, or Lucky Red Carpet cans for $5.

Who falls for these ripoffs? Kids, mostly. Which is too bad, because they could easily trade for these very same cans 1-for-1 at any trade session! (Or through the mail.)

All of which has infuriated Bob McClure so much that he threatens to start a brand new organization called the SPBCCEFBOH. Which means, in case it wasn't immediately clear to you, the Society for the Preservation of Beer Can Collecting and Elimination of the Fast Buck Operators from the Hobby.

Henry Herbst would be the first to join. "An epidemic of Purveyors of Painted Tin and Aluminum is sweeping the country," he says. "The result is that many unsuspecting collectors are spending needless amounts of money on overvalued objects." Herbst also fears that if buying and selling of cans becomes the rule rather than the exception "a fantastic hobby will be turned into a help-yourself-and-to-heck-with-everyone-else affair."

McClure calls them Fast Buck Operators. Herbst calls them Purveyors of Painted Tin and Aluminum. Whatever they are—and whoever they are—these can merchants *do* exist. There are probably half a dozen who are earning handsome livings trafficking in current and recently-obsolete cans.

Clark Secrest isn't sure they're going to be going away soon, either. Secrest is the founder of the Tin Container Collectors Association, a club that deals—usually with money—in antique tobacco and coffee cans, etc. "By popularizing the hobby," he says, "the BCCA must be held accountable for the existence of the can merchant. If there were no BCCA, there would be far fewer collectors, much less exchange of information—and no dealers. Whenever the number of people collecting something grows to a sizeable number, inevitably dealers in that something come to be."

Secrest also points out that the only reason there are dealers is because they have something to offer that the collector can't find elsewhere.

BCCAer Gil Brennell just shakes his head at this. "Why do we need dealers to find cans for us? We have thousands of collectors we can trade with for those same cans!"

Even Secrest—who is not at all sure buying and selling won't inevitably take over beer can collecting—agrees on this point. "Dealers will not be able to survive unless they can come up with beer cans that somehow escape collectors. If they can't it will be curtains for the dealers."

Keeping beer can collecting a price-less hobby may be difficult, but it is not impossible if three things happen.

1. There has to be more communication than there has been between collectors from all parts of the country. This will stimulate trading in new currents as they come on the market and will shut out the dealer in currents.

2. There has to be a better understanding of can values for all grades of cans. Both to facilitate trading of tough and semi-tough cans and to keep these cans from becoming overvalued. This will cause the dealer who offers exorbitantly-priced cans to be laughed into oblivion. (Or, better yet, ignored to death.)

3. There has to be a real shockeroo sent through the beer can world. Proof that beer can values are fickle will hold values down where they belong.

What kind of shockeroo? Actually, there have been two already, but they apparently weren't enough to do the trick. First, there was the reissuing of Olde Frothingslosh—and the instant plummeting of the value of that can (whether expressed in trading power or in dollars).

Then there was the instant replay of the Frothingslosh phenomenon with King Snedley's. By now, you'd think the collecting world would be wary of overvaluing certain cans. But no.

It should be, especially with reproductions headed our way. (A good question: If you were in the reproduction business, which beer cans would you decide to reproduce? *Right*.)

Then there's the case (actually, 161 cases) of Rich Lehecka and Larry Klimes to consider. They found a shed full of conetops—3,875, to be exact. All mint. Five different brands. Which would make you feel a bit down in the wallet if you'd invested cash in any of those particular brands. And big finds will happen again. And again.

And who knows what brands will be suddenly de-valuated?

Rumors, rumors, rumors. Did you hear that 30 cases of James Bond 007's have been found? Or that 50 cases of mint Eastside cones just showed up? Did anybody tell you about that pallet of Soul—? (It's not impossible. Not a bit. After all, they *did* make over a million cans of the stuff.)

Whenever somebody tells you a can is worth $5 or more, think about the reissuing possibilities. Then think about Rich and Larry's big find.

There. That should do it. That should keep things in perspective.

And it's important that collectors *do* work together to keep inflation from running up the value of beer cans because that just gets the Fast Buck Operators into the act.

And when the barter part of our hobby is gone, the worst part—the money part—will take over. Then it won't be a hobby any more.

It will be a business.

16

Trading—and getting more than you gave

> "The best thing about trading cans is you're helping somebody else's collection while you're helping your own."
>
> Morris McPherson, Beer Can Collector
> Sycamore, Illinois

In any swap of beer cans between consenting adults, both parties wind up with more than they gave. That's the beauty of this important aspect of the hobby. A can that's common where you live may be a prize to another collector across the country. And vice versa.

Face-to-face trading is the easiest to do, of course. Because then you can pick through the other guy's traders, check out the condition of the ones you're interested in, spot minor label variations—and then negotiate. Negotiation—that's the tricky part, because you may not know the relative value of the cans involved.

If both of you are candid about your cans and reveal everything you know about them, there's usually no problem in putting together a deal. Good sportsmanship is important. As is not trying to wring everything you

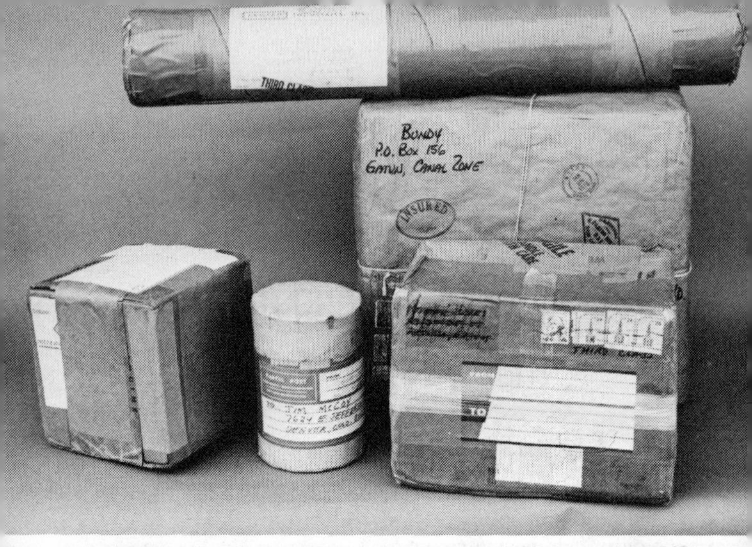

PROPERLY WRAPPED, beer cans survive anything the Post Office can throw at them. Or stack on top of them. By-mail trading is a popular way to build your collection.

possibly can out of every can you trade. If you build a reputation as a tough trader—and some collectors have—you'll find fewer people who'll want to go through the agony of trading with you.

Although face-to-face trading is fast and easy, there are many collectors you'll never meet in person who have some find cans to trade with. That's why trading-by-mail has become so popular. (If the U.S. Postal Service only knew how many of those packages were just empty beer cans!)

The first thing you have to do if you're going to conduct by-mail trading is to make yourself a trading list. This list gives the brand names of all the cans you're putting up for trade. It should also name the brewery and city of origin printed on each can, give any additional description necessary to differentiate it from another can of the same brand, and indicate its condition.

Nothing infuriates (and disappoints) a collector more than to open a package and discover a dumper when he

BEER CAN TRADING LIST PLEASE RETURN (UNMARKED)

Jim McCoy/7624 E. Jefferson Dr./Denver, Colo. 80237 BCCA #136

BRAND	TYPE	BREWERY	SIZE	METAL/TOP	COLORS NAME/MAIN/OTHER	REMARKS
Bavarian Club	B	J. Huber	12	St/Tab	Wt/Wt/Yw-Be-Rd-Sr	
Berghoff 1887 OBS.	B	Walter (Colo)	12	St/Tab	Bk/Wt/Rd-Gd	
Berghoff 1887 Draft OBS.	DB	Walter (Colo)	12	St/Tab	Wt/Bn/Gd-Bk	
Brown Derby Lager	B	Walter (Colo)	12	St/Tab	Wt-Bn/Wt/Rd-Bn-Gd	
Buckhorn	B	Buckhorn	12	St/Tab	Wt/Gn/Rd-Bk	
Cold Spring	B	Cold Spring	12	St/Tab	Wt/Wt/Be-Sr	
Colorado Gold Label	B	Walter (Colo)	12	St/Tab	Wt/Be-Gn/Rd-Gd	Mountain scene
Colt 45 Stout	ML	National	8/12/16	St/Tab	Be-Rd/Wt/Rd-Gd	
Coors	B	Coors	7/12/16	Al/Tab	Bk-Sr/Yw/Sr-Rd-Bc	
~~Country Club~~	ML	Pearl	12	St/Tab	Rd/Wt/Gd-Be	White XXX
Crystal Colorado	B	Walter (Colo)	12	St/Tab	Wt-Sr/Wt/Be-Gd-Rd	
~~Draft Beer~~	DB	National	12	St/Tab	Rd/Wt/Bn-Yw-Sr	
Drummond Bros.	B	Drummond	12	Al/Tab	Gd-Bk/Sr/Wt-Bk	
[illegible]	B	Duquesne	12	St/Tab	Bk-Rd/Wt/Gd-Rd-Bk	
[illegible]	B	Associated	12	St/Tab	Rd-Yw/Wt/Be-Gn-Bn	
Falls City	B	Falls City	12	St/Tab	Wt/Wt-Rd/Gd	
Gablinger's	B	Forrest	12	St/Tab	Bn-Gd/Bn/Tn	"99 Calories"
Goetz	B	Pearl	12	St/Tab	Wt/Wt/Be-Sr	Scratched
Gold Label	B	Walter (Colo)	12	St/Flat	Rd/Gd/Wt-Rd	
Grain Belt	B	Grain Belt	12	St/Tab	Wt/Gd/Rd-Wt	
Great Falls Select	B	Blitz-Weinhard	12	St/Tab	Wt/Wt/Rd-Gd	
Hamm's	DB	Hamm	12	St/Tab	Wt/Sr/Be	Barrel shape
Hamm's Preferred Stock	B	Hamm	12	St/Tab	Bk/Wt/Bk-Rd	
Harlequin Old Fashioned Dark	B	Walter (Colo)	12	St/Tab	Rd-Bk/Yw/Gd-Wt	
Harlequin Old Fashioned Dark	B	Walter (Colo)	12	St/Tab	Rd-Bk/Yw/Gd-Wt	Oklahoma stamp
Heidelberg	B	Carling	12	Al/Tab	Rd/Gd/Rd-Wt	
Heidel Brau Pilsner	B	Heidel Brau	12	St/Tab	Wt/Wt/Rd-Gy	
Heileman's Old Style	B	G. Heileman	12	St/Tab	Wt/Wt/Be-Rd-Yw-Bn	
Heileman's Special Export	B	G. Heileman	12	St/Tab	Bk-Rd/Wt/Gn	
Hi Brau Premium	B	J. Huber	12	St/Tab	Wt/Gd/Bk-Rd	Shield & lion
Huber Premium	B	J. Huber	12	St/Tab	Wt/Wt/Rd-Gd-Bk	
Hynne	B	Walter (Colo)	12	St/Tab	Rd-Bk/Yw/Sr-Rd	
Kassel OBS.	B	Jackson	12	St/Tab	Bk/Gd/Bk-Rd	
Kingsbury	B	Kingsbury	12	St/Tab	Rd/Wt/Rd-Be-Gd	

A GOOD TRADING LIST makes it easy to know what you're trading for. All the pertinent information about each can is included in this collector's list of traders.

had visions of a mint can dancing in his head. The generally-accepted rule is that if no mention of condition is given, one can assume the can is in very good shape. If a can you're offering is full of machine gun holes, was run over by a tank, or spent the last twenty years at the bottom of a pond—be sure to mention it.

A lot of collectors use a grading system developed by Bob Myers as a kind of shorthand to indicate a can's condition:

GRADE 1: Top quality. No dents, rust spots or other easily noticeable imperfections. An otherwise perfect can may be considered Grade 1 even if it has slight scratches providing they are noted in your trading list.

GRADE 2: An excellent display can although it may have small scratches, dents or rust spots. The unpainted top or bottom may be rusty, but the sides are basically in good, clean condition.

GRADE 3: A satisfactory display can, but it has easily seen scratches, dents or faded/rusty areas. To be Grade 3, a can must still retain a good degree of its original appearance on *all* sides.

GRADE 4: A poor display can with major flaws such as large faded/rusty areas, many scratches or dents. At least one *full* side must qualify as being of Grade 3 quality, thus making the can suitable for display as a filler.

GRADE 5: Major imperfections across all sides of the can make it difficult to read or see the original colors and design.

1- BASIC HAMM'S "PREFERRED STOCK" CAN PRE-1950. GOLD BACKGRND. RED RIBBON & STARS & BLUE LETTERING.

2- BASIC SCHMIDT'S "CITY CLUB" GOLD BACKGROUND BLACK LETTERING EXCEPT "CITY CLUB" IN RED (PRE-1950).

3- BASIC OLD PABST "BLUE RIBBON" GREY HORIZONTAL BARS, GOLD RHOMBOID BORDERS, BLUE LETTERING & RIBBON.

4. STANDARD GRAIN BELT SPOUT ORANGE "WAVY" BACKGRND. ORANGE DIAMOND BROWN LETTERING EXCEPT GRAIN BELT WHICH IS OFFSET. YELLOW BOTTLE CAP IS UNDER DIAMOND. PRE-1950.

5. PABST "EXPORT" 1935-38 PRE-PATENT (LIMITED NUMBERS AVAILABLE) GREY BCKGRND. RED PABST & SEAL W/BLUE HOP LEAF & BLUE BCKGRND AROUND GREY "BEER." PABST'S FIRST CAN. (3 ONLY)

CLARK L. WALKER

MAKE AN OFFER: 1008 10TH AVE NW #C MINOT ND 58701

AN ILLUSTRATED TRADING LIST is helpful when there are label designs involved that are hard to accurately describe.

Anything you can add to more clearly describe the cans you're offering (has no top, is full of axle grease, etc.) should be mentioned. Some collectors are reluctant to tell all because they're afraid they'll lose a trade. Better that than to lose your reputation.

Some brands have produced cans with such a wide range of label variations that it's difficult to tell which a trading list refers to—unless a breakdown of the colors in the label design is given. Here, then, is a system for describing a can's color scheme which is used by many can collectors:

First, list the color of the letters in the brand name. (If there are two colors, list the dominant one first and the secondary one second.)

Next, give the color surrounding the name (or the background on which the name is printed).

Then, give the next most prominent color.

Finally, give the next most prominent color.

Using this system, the colors of the current Coors can would be described as BLACK-SILVER-BUFF/ RED/BLACK.

Of course, there's a neat way you can eliminate a whole lot of the uncertainty concerning the specific label design of a can or its condition. Just send pictures. If they're color pictures, you've taken care of the pesky color problem, too. A number of collectors fill a flat (empty beer case) with cans lying on their sides, load up the old Polaroid with color film, and fire away. It's a proven fact that when you send out photographs of your traders that show all the details of a can's design and condition you make more trades. A top-loading Xerox (or similar) machine can also be used to give that guy across the country a better idea of what the can you've got to trade looks like. No camera? No Xerox? Then why not *draw* a picture of any cans you're afraid won't be fully appreciated without some kind of visual representation? A lot of guys do it. If it didn't work they wouldn't bother.

No camera? No Xerox? And absolutely no drawing ability? No problem. Just trade on approval. Send him a box of your best, let him pick what he needs, then negoiate the cans he'll send you in exchange through correspondence. Eventually, you'll get back the traders he didn't want—plus the cans he swapped you for the cans he did want.

One warning: Don't start shipping off boxes of cans on approval without first establishing contact by mail and getting the permission of the recipient-to-be. Not everybody wants to trade on approval.

A few more things to think about before you put together your list of traders: Separate your domestic and foreign cans. Consider *not* listing your cans in alphabetical order (this will cause the reader to study your list more carefully and this may lead him to discover some of your jewels he might have overlooked if he'd just glanced over it). Indicate obsolete cans with an asterisk.

And be sure to make note of any limitations you put on your trading. If you will only trade obsoletes for other obsoletes, say so. If you wouldn't consider trading that gallon for anything but another gallon, let him know. If you consider your foreign cans to be worth two current U.S. cans, mention it. Matter of fact, you might consider prefacing your trading list with a statement as to what kinds of cans you're looking for (and not looking for) and your general philosophy of trading.

Consider adopting a policy of permitting return privileges. Sort of like offering a cans-back guarantee. The collectors who have this policy—and mention it on their trading lists—generally make more trades. And have more guys around the country talking about what a pleasure it is to trade with good old so-and-so. Not a bad reputation to have.

Another thing: Don't send copies of your trading list to too many collectors all at once. You may find yourself suddenly traded out of one thing or another and that'll have you spending a lot of time writing apologies.

When it comes to saving time, of course, there are lots of ways to avoid having to laboriously copy your trading list over and over. Typing it—with lots of carbons—is one way. Those collectors who have access to mimeograph machines or office photocopiers usually make good use of them, too.

A couple of notes on trading etiquette. A preliminary offer of cans to trade should be ended with "*if* you want these, let me know." (Then, if he doesn't want them, he doesn't feel obligated to spend time and money writing you a note just to say, "No, thanks.") If, on the other hand, someone is holding cans for you, waiting to see if a trade can be worked out, you'd be very remiss if you didn't answer. In any case, when a reply is called for, make it a prompt one. It can keep you from losing cans you could've traded for. (The good ones generally gc pretty fast.)

Some collectors want you to return their trading lists, other don't. Make sure you indicate—one way or the other—on your trading list. And, when you receive a list that is to be returned, have the courtesy to comply.

Who pays the postage? A few collectors have gotten into the S.A.S.E. (self-addressed, stamped envelope) thing. Some ask for them, others supply them without being asked to. And sometimes postage becomes an issue when a large, expensive shipment is involved. Generally, though, most collectors just pay the postage for anything they send and let it go at that.

When some collectors spot something on a trading list they *just have to have*, they're so afraid that somebody will get there first that they drop everything and call. This might seem pretty extravagant considering they're just calling about an empty beer can, but if you use evening rates and keep your call down to the minimum, it's not a bad method.

One quick call can take the place of weeks of back-and-forth negotiating by mail.

Now for the critical part—getting those priceless Grade 1 jewels to the guy you traded with without having them arrive in Grade 5 condition.

BCCA member John MacIntyre of Holmes, Pennsylvania is a postal inspector. Here's what he says parcels—all parcels—go through between here and there: "They're generally put in sacks which can weigh up to seventy pounds when full. These sacks are then stacked in trucks and railcars—which can result in several hundred pounds resting on those sacks on the bottom. The vehicles, of course, shift back and forth as they move. To prevent damage, pack your cans in good, heavy, corregated cartons with balled-up newspaper in the empty spaces between, over, and under the cans."

Collector Don Bannon suggests shipping cans in the

hard cardboard cylinders carpet comes wrapped around. "Visit your local carpet store," he says. "You can cut those 13-foot tubes into any length you need." Bannon also suggests placing wads of paper between each can and at both ends, then taping cardboard caps on the ends.

"Beer cans are surprisingly fragile," warns collector Max Robb, the postmaster of Central City, Colorado. "Especially the aluminum ones. Shoe boxes and cereal boxes don't offer nearly enough protection. And those cans that Pringles Newfangled Potato Chips come in aren't nearly fangled enough to do the job, either."

Two of America's biggest (and best) mailers of beer cans are Ken and Lois Hiestand of York, Pennsylvania. The Hiestands are both BCCA members, and they've got a rather unique approach to the hobby. Often. they don't even bother with trading. They just pack up boxes of cans truckdriver Ken has run across (not literally, thank goodness!) in his travels and ship them to their collector friends around the country. Completely unsolicited! And you know what? Every so often the postman shows up with a couple of boxes for the Hiestands, too.

Trading beer cans is fun, but what the Hiestands do just might be more fun.

17

Relative rareness ratings

"I'll trade anybody a Schmidt's of Philadelphia for a Sierra."

Dom Mazzeo, Beer Can Collector
Philadelphia, Pennsylvania

The hardest part of making a trade for a beer can you don't have is knowing what it's "worth." Knowing the value of the traders you *do* have is nearly as difficult.

Is Mazzeo's Schmidt's-for-a-Sierra offer a good even-up trade? Probably not. But how can you know for sure? After many hours of trading and hanging around trade sessions, a collector gets a handle on the relative availability and desirability of the cans in his trading stock. Similarly, he begins to learn how tough the cans he hankers for are.

Still, even after years of trading, a tremendous amount of disagreement as to the toughness of thousands of cans remains.

To provide a guide to trading values, some people rate cans in terms of how many current cans (cans that are now available on the market) they are worth. Some say, for example, that an older Schmidt's of Philadelphia is a 2 for 1 trader; it's supposed to be worth two currents. A Sierra, on the other hand, is a 5 for 1 can.

The problem with this system is obvious: Nobody in their right mind would trade a Sierra for five currents. Nor would they trade one for *thirty* current cans. They'd want to get another tough can for it. Current cans are excellent cans for trading for other current cans and often can be traded on a multiple-for-1 basis for recent obsolete cans, but to use currents to set values for the older, most-wanted cans just isn't realistic.

For this reason, a new system—Relative Rareness Ratings—has been developed.

Five experienced collectors, all currently active in the trading market, independently rated the cans listed below on a scale of one to twenty-five. The average ratings they gave these cans express how tough they estimate these cans to be. The higher the Relative Rareness Rating, the more difficult they believe a can is to come by.

A can with an RRR of 1 would be very common, extremely easy to find. An RRR of 5 indicates that a can is slightly tough. RRR 10 means that a can is medium tough. (You're mighty glad you found one!) A can rated as RRR 15 is a very tough can. (You can expect to give a lot to get one . . . or get a lot if you have an extra for trade.) The cans listed at RRR 20 or above are very , *very* tough cans. A perfect score of RRR 25 would only go to a can that is one of the world's rarest two-dozen cans.

The collectors who supplied the Relative Rareness Ratings did not attempt to rate all the beer cans ever made. Instead, they limited themselves to only U.S. 12 ounce flattops and pull-tabs (no cones, odd-sizes or gallons). They rated only obsoletes, since currents would automatically be RRR 1 or close to it. They made the assumption that each can was in top condition.

The cans they chose to rate are representative of the cans a collector may encounter no matter in what part of the country he lives. Some are fairly easy to come by. Others are mighty scarce. The rest are somewhere in between.

Here, then, is the Relative Rareness Rating of over 200 beer cans in the opinion of five collectors with 1,500, 2,000, 2,200, 2,300 and 4,600 cans in their collections. (One warning: The reissue of any one of these cans or the discovery of a large cache of them would immediately lower its RRR.)

Brand Name	Brewer	Principal Location of Brewer	BCCA Code Number	Relative Rareness Rating
ABC	Maier	Los Angeles, CA	MR 3	9.8
Acme	Acme	Los Angeles, CA	ACM 1	14.2
All Grain	Storz	Omaha, NB	STZ 10	13.8
Amber Brau	Maier	Los Angeles, CA	MR 4	6.6
A-1	National	Phoenix, AZ	CAR 1	9.4
Aristocrat	Mountain	Denver, CO	MT 4	17.8
Ballantine	Ballantine	Newark, NJ	BAL 4	13.0
Barbarossa	Atlantic	South Bend, IN	ALT 7	11.8
Becker's Best	Becker	Ogden, UT	BKR 9	17.2
Becker's Uinta Club	Becker	Ogden, UT	BKR 7	15.6
Berghoff 1887	Walter	Pueblo, CO	WLC 6	4.8
Big Apple	Waukee	Hammonton, NJ	WK 1	14.2
Big Mac	Menominee-Marinette	Menominee, MI	MM 3	17.4
Black Pride	West Bend Lithia	West Bend, WI	WBL 1	4.6
Blatz	Blatz	Milwaukee, WI	BL 2	12.2
Blue 'N Gold	North Bay	Santa Rosa, CA	NB 1	20.6
Bohack	Richards	Newark, NJ	RD 2	5.0
Bonanza	Garden State	Hammonton, NJ	GSI	7.0
Brew 66	Sicks	Seattle, WA	SK 4	18.4
Brown Derby	L.A.	Los Angeles, CA	LA 3	13.2
Budweiser	Anheuser-Busch	St. Louis, MO	AB 2	10.8
Budweiser Bock	Anheuser-Busch	St. Louis, MO	AB 16	24.4
Bulldog	Grace Brothers	Santa Rosa, CA	GB 3	10.4
Bullfrog	Monarch	Chicago, IL	MON 3	10.4
Burgermeister	San Francisco	San Francisco, CA	SF 1	13.2
Busch Lager	Anheuser-Busch	St. Louis, MO	AB 6	18.0
Busch Bavarian	Anheuser-Busch	St. Louis, MO	AB 101	6.6
Canadian Ace	Canadian Ace	Chicago, IL	CAN 3	8.0
Cardinal	Cardinal	St. Charles, MO	CD 1	4.4
Clear Lake	Grace Bros.	Santa Rosa, CA	GB 4	17.4
Clipper Pale	Grace Bros.	Los Angeles, CA	GB 54	22.0
Colorado Imperial	Walter	Pueblo, CO	WLC 9	11.8

Brand Name	Brewer	Principal Location of Brewer	BCCA Code Number	Relative Rareness Rating
Congress	Haberle	Syracuse, NY	HC 13	17.2
Coors	Coors	Golden, CO	COR 5	16.2
Crystal Colorado	Walter	Pueblo, CO	WLC 22	4.0
Dakota	Dakota	Bismarck, ND	DAK 1	13.2
Dawson	Dawson	Willimasett, MA	DAW 1	6.6
Dixie	Dixie	New Orleans, LA	DX 2	15.2
Dodger	Maier	Los Angeles, CA	MR 6	13.0
Drewrys	Drewrys	South Bend, IN	DR 7	9.4
Du Bois Budweiser	Du Bois	Du Bois, PA	DU 3	11.0
Duke	Duquesne	Pittsburgh, PA	DUQ 1	8.6
Dunk's	Duncan	Auburndale, FL	DUN 1	4.8
Dutch Lunch	Grace Bros.	Los Angeles, CA	GB 44	18.6
Eastern	Atlas	Chicago, IL	ATS 3	13.0
Eastern	Drewrys	South Bend, IN	DR 48	15.8
Eastside	Los Angeles	Los Angeles, CA	LA 1	13.8
Edelweiss	Schoen-Edelweiss	Chicago, IL	SE 9	7.6
Einbock	Walter	Pueblo, CO	WLC 18	5.2
Encore	Schlitz	Milwaukee, WI	SZ 2	4.8
Esslinger	Esslinger's	Philadelphia, PA	ES 3	15.2
18-K	Oconto	Oconto, WI	OCT 3	13.2
Fabacher Brau	Jackson	New Orleans, LA	JAC 1	4.2
Falstaff	Falstaff	St. Louis, MO	FA 4	9.0
Fehr's X/L	Fehr	Cincinnati, OH	FHR 1	7.8
F & G	Drewrys	South Bend, IN	DR 30	11.2
Fitz	Fitzgerald Bros.	Troy, NY	FZ 1	10.8
500 Ale	Cook	Evansville, IN	COK 1	16.4
Food Fair	Yuengling	Pottsville, PA	YU 1	17.2
49er	Atlas	Chicago, IL	ATS 4	20.6
Fox Head 400	Fox Head	Waukesha, WI	FXH 5	15.0
Frankenmuth Bock	Frankenmuth	South Bend, IN	FRM 1	10.2
Gambrinus	Wagner	Columbus, OH	WAG 2	4.8
Gamecock	Cumberland	Cumberland, MD	CU 3	19.2
G B	Griesedieck Bros.	St. Louis, MO	GR 1	10.2
G.E.M.	Fuhrmann & Schmidt	Shamokin, PA	FS 3	11.6
Genesee Cream Ale	Genesee	Rochester, NY	GSE 8	8.2
Gettelman	Gettelman	Milwaukee, WI	GET 4	14.4
Glacier	Maier	Los Angeles, CA	MR 9	7.2
Gluek's	Gluek	Minneapolis, MN	GLK 2	10.6
Goebel	Goebel	Detroit, MI	GBL 10	13.0

Brand Name	Brewer	Principal Location of Brewer	BCCA Code Number	Relative Rareness Rating
Golden Gate	Maier	Los Angeles, CA	MR 58	13.6
Golden Lager	Burgermeister	San Francisco, CA	BGM 1	4.4
Gold Coast	Drewrys Ltd.	South Bend, IN	DR 22	14.4
Gold Label	Walter	Pueblo, CO	WLC 19	7.0
Grain Belt	Minneapolis	Minneapolis, MN	MN 7	12.4
Grand Prize	Gulf	Houston, TX	GUL 3	14.6
Great Falls	Great Falls	Great Falls, MT	GF 1	13.4
Great Lakes	Schoen-Edelweiss	Chicago, IL	SE 4	8.4
Gretz	Wm. Gretz	Philadelphia, PA	GRZ 7	19.6
Gretz	Wm. Gretz	Philadelphia, PA	GRZ 17	21.0
Guiness's	Goebel	Detroit, MI	GBL 13	19.6
Hal's	Hals	Baltimore, MD	Hal 1	18.6
Hamm's	Hamm	St. Paul, MN	HM 4	10.4
Happy Hops	Grace Bros.	Santa Rosa, CA	GB 26	20.4
Hapsburg	Best	Chicago, IL	BST 3	15.4
Harlequin	Walter	Pueblo, CO	WLC 10	3.6
Hartz Western Style	Silver Springs	Tacoma, WA	SS 1	13.2
Harvard Ale	Harvard	Lowell, MA	HVD 4	16.6
Heidelberg	Carling	Tacoma, WA	CAR 8	8.6
Highlander	Missoula	Missoula, MT	MSA 1	13.8
Home Bock	Atlas	Chicago, IL	AT 5	13.6
Hop'n Gator	Pittsburgh	Pittsburgh, PA	PIT 3	5.2
Hull's Cream Ale	Hull	New Haven, CT	HUL 4	16.4
Hyde Park 75	Hyde Park	St. Louis, MO	HP 1	16.2
India	San Juan	Hammonton, NJ	CSJ 2	10.4
Iron City	Pittsburgh	Pittsburgh, PA	PIT 52	9.2
Iroquois Ale	Iroquois	Buffalo, NY	IRQ 5	11.6
Jaguar	Jaguar	Rochester, NY	JAG 1	12.8
Jamaica Sun	Jamaica	Reading, PA	JAM 1	6.8
Jax	Jackson	New Orleans, LA	JAC 4	6.6
KC's Best	Storz	Omaha, NE	Stz 11	5.4
Karl's	Grace Bros.	Santa Rosa, CA	GB 28	13.2
Katz	Associated	Evansville, IN	AS 11	6.4
Kentucky Malt Liquor	Fehr	Louisville, KY	FHR 5	16.6
King Cole	Maier	Los Angeles, CA	MR 49	21.4
King's	King's	Chicago, IL	KG 1	14.4
Kingsbury Draft	Heileman	Sheboygan, WI	HMN 18	9.4
Knickerbocker Dark	Ruppert	New York, NY	RP 13	16.4
Koch's	Koch	Dunkirk, NY	KCH 2	10.4

Brand Name	Brewer	Principal Location of Brewer	BCCA Code Number	Relative Rareness Rating
Kool	Grace Bros.	Los Angeles, CA	GB 47	22.0
Krueger	Krueger	Newark, NJ	KGR 3	15.2
L&M	Maier	Los Angeles, CA	MR 50	17.8
Lebanon Valley	Eagle	Catasauqua, PA	EGL 1	19.2
Land of Lakes	Pilsen	Chicago, IL	PLN 1	12.2
Lime Lager	Lone Star	San Antonio, TX	LS 3	10.2
Linden Light	Colonial	Hammonton, NJ	CLL 7	18.4
Lone Star	Lone Star	San Antonio, TX	LS9	9.4
Lubeck	Lubeck	Chicago, IL	LUB 1	16.6
Lucky Lager	Lucky Lager	San Francisco, CA	LKY 4	12.8
Malt Duck (Grape)	National	Baltimore, MD	NAT 11	9.6
Manhattan	Manhattan	Chicago, IL	MAN 8	17.2
Matterhorn	Burgermeister	San Francisco, CA	BGM 4	9.2
McLab	Drewrys	South Bend, IN	DR 26	15.8
Meister Brau	Peter Hand	Chicago, IL	HDP 19	14.8
Metz	Metz	Omaha, NE	MTZ 1	14.2
Mile Hi	Tivoli	Denver, CO	TIV 11	17.2
Miller	Miller	Milwaukee, WI	MLR 17	17.2
Milwaukee Extra	Miller	Milwaukee, WI	MLR 9	6.2
Monticello	Monticello	Norfolk, VA	MCL 1	15.8
Muehlebach	Muehlebach	Kansas City, MO	MU 1	17.0
Narragansett	Narragansett	Cranston, RI	NRT 8	10.8
National Bohemian	National	Baltimore, MD	NAT 13	6.6
Nectar	Ambrosia	Chicago, IL	AMB 1	14.8
9-0-5	9-0-5	Chicago, IL	NOF 1	8.2
Norvic	DuBois	DuBois, PA	DU 7	9.2
007	National	Baltimore, MD	NAT 33	21.2
Oertels 92	Oertel	Louisville, KY	OER 4	10.0
Olbrau	Metropolis	Trenton, NJ	MET 5	17.4
Old Bru	Hamm	San Francisco, CA	HM 12	10.4
Old Crown Ale	Centlivre	Ft. Wayne, IN	CE 1	17.8
Old Dutch	Krantz	Findlay, OH	KRZ 1	14.4
Old Georgetown	Christian Heunrich	Washington, D.C.	HRH 1	18.6
Old German	Colonial	Hammonton, NJ	CLL 13	17.6
Old Gibraltar	Maier	Los Angeles, CA	MR 45	16.6
Old Milwaukee	Schlitz	Milwaukee, WI	SZ 5	10.4
Old Ranger	Hornell	Hornell, NY	HNL 1	22.2
Old Style	Heileman	LaCrosse, WI	HMN 21	4.2
Old Tavern	Warsaw	Warsaw, IL	WR 3	6.6

Brand Name	Brewer	Principal Location of Brewer	BCCA Code Number	Relative Rareness Rating
102 Dark	Maier	Los Angeles, CA	MR 28	14.4
Orbit	Orbit	Tampa, FL	OBT 1	6.2
Ortlieb's	Ortlieb	Philadelphia, PA	ORT 6	10.4
Ox Bow	Walter	Pueblo, CO	WLC 23	16.4
Pabst Ale	Pabst	Milwaukee, WI	PBT 24	16.2
Pabst	Pabst	Milwaukee, WI	PP 3	13.6
Patrick Henry M.L.	Fox	Chicago, IL	FXP 5	22.4
Paul Bunyan	Wisconsin	Waukesha, WI	WS 3	15.2
Pearl	Pearl	San Antonio, TX	PRL 8	9.2
Pearl Draft	Pearl	San Antonio, TX	PRL 9	10.8
Peoples	Peoples	Oshkosh, WI	PLS 1	4.4
Pfeiffer's	Pfeiffer	Detroit, MI	PFR 8	15.2
Piels	Piel Bros.	Brooklyn, NY	PL 2	4.0
Pikes Peak Ale	Walter	Pueblo, CO	WLC 25	12.8
Pilsengold	San Francisco	San Francisco, CA	SF 3	18.2
Playmate M.L.	Sunshine	Reading, PA	SUN 3	22.4
P.O.N.	Feigenspan	Newark, NJ	FGS 4	20.0
Primo	Schlitz	Milwaukee, WI	SZ 10	3.6
Pub M.L.	Home	Richmond, VA	HME 1	12.8
Rahr's All Star	Rahr	Green Bay, WI	RHR 1	12.8
Rainier	Rainier	Seattle, WA	RNR 3	8.6
Red Lion	Red Lion	Cincinnati, OH	RLI 1	8.2
Redtop	Redtop	Cincinnati, OH	RT 2	15.4
Regal Select Draft	Maier	Los Angeles, CA	MR 52	12.6
Regency	Maier	Los Angeles, CA	MR 70	11.8
Renner	Renner	Ft. Wayne, IN	Rnn 2	7.0
Rheingold Ale	Liebmann	New York, NY	LBM 2	10.6
Rheingold	Liebmann	New York, NY	LBM 7	19.0
Riviera	Atlantic	Chicago, IL	ATL 12	18.8
Rolling Rock	Latrobe	Latrobe, PA	LTB 3	12.6
Royal Amber	Wiedemann	New Port, KY	WMN 11	16.0
Royal 58	Duluth	Duluth, MN	DL 1	11.0
Ruppert	Jacob Ruppert	New York, NY	RP 18	14.0
Santa Fe	Maier	Los Angeles, CA	MR 55	16.4
Schaefer	Schaefer	Brooklyn, NY	SCH 8	15.2
Schlitz	Schlitz	Milwaukee, WI	SZ 36	10.4
Schmidt's City Club	Schmidt	St. Paul, MN	STJ 4	14.4
Schmidt's Bock	Schmidt	Philadelphia, PA	STC 19	16.2
Schoenling Bock	Schoenling	Cincinnati, OH	SG 1	20.6

Brand Name	Brewer	Principal Location of Brewer	BCCA Code Number	Relative Rareness Rating
Schoen's	Wausau	Wausau, WI	WSU 1	15.6
Sebewaing	Sebewaing	Sebewaing, MI	SW 2	17.8
Seven-Eleven	Lexington	Newark, NJ	LEX 1	9.2
Silver Cream	Menominee-Marinette	Menominee, MI	MM 1	15.2
Simon Pure	Simon	Buffalo, NY	SMN 1	14.2
Ski Country	Walter	Pueblo, CO	WLC 15	7.4
Soul Mellow Yellow	Maier	Los Angeles, CA	MR 36	23.4
Spur M.L.	Sicks Rainier	Seattle, WA	SR 6	15.0
Stallion XII M.L.	Gold Medal	Wilkes-Barre, PA	GM 2	11.6
Storz Draft	Storz	Omaha, NE	STZ 7	12.0
Sunshine	Sunshine	Reading, PA	SUN 2	11.8
Tex	Jackson	New Orleans, LA	JAC 2	5.6
Tomahawk Ale	International	Buffalo, NY	INT 2	14.8
Top Hat	Tivoli	Denver, CO	TIV 13	21.4
University Club M.L.	Gettleman	Milwaukee, WI	GET 1	14.6
Utica Club Bock	West End	Utica, NY	WE 18	19.4
Velvet Glove, M.L.	Hamm	St. Paul, MN	HM 11	12.2
Weisbrod	Old Dutch	Allentown, PA	OD 6	10.0
Whale's White Ale	National	Baltimore, MD	NAT 4	12.2
Wilco	Colonial	Hammonton, NJ	CLL 6	8.0
Wisconsin Club	Huber	Monroe, WI	HBR 4	6.6
Yuengling	Yuengling	Pottsville, PA	YU 3	7.8
Yusay	Pilsen	Chicago, IL	PLN 4	15.8

18

You can't remember all your cans, you know

> "If I told the guys I work with at the bank that I had a collection of empty beer cans, they'd laugh right in my face. So I tell them I collect metallic malted beverage containers."
>
> Stuart Peterson, Beer Can Collector
> St. Paul, Minnesota

After you've got a couple of hundred different cans in your collection, you start to forget . . . "Do I have that Genesee Ale or does mine say Genesee *Light* Cream Ale?" "Is the schooner of beer on the front of that Bohack in my collection a drawing or a photograph?" "Does my Milwaukee's Best have one tankard on the label—or two?"

When this kind of confusion sets in, one of two things happens. Either you trade for a can not knowing you've already got it in your collection (bad). Or you *don't* trade for a can you should have (worse).

When you're trading in close proximity to your collection, of course, this is no problem. But when you're at a trading session or at a Canvention, you're in deep trouble.

That's why many collectors have worked out systems to keep track of what cans they have. Some are elaborate, some are simple. All help.

The most obvious method of keeping track of what you've got up on the wall at home is to make a list. An alternative to this, used by many collectors, is to put a

ROY FUERHERM #3
100 CHATHAM PK
PITTSBURGH, PA
SEPTEMBER 30, 1

ALL BEER CANS ARE 12 OUNCE EMPTY AND IN PERFECT CONDITION. MOST CAN
BOTTOM OPENED. ALL CANS ARE 1 FOR 1 UNLESS STATED OTHERWISE.

BEER CAN NAME	BREWERY	
A-1 PREMIUM (OLD)	NATIONAL BREWING	
ABC	GARDEN STATE BREWING	
ABC ALE	GARDEN STATE BREWING	
ACME	ACME BREWING	
ALTES	NATIONAL BREWING	
B BAVARIAN	MOUNT CARBON BREWING	
BALLANTINE BOCK	FALSTAFF BREWING	
BALLANTINE DRAFT	FALSTAFF BREWING	
BARTELS	LION BREWING	
BERGHEIM	BERGHEIM BREWING	
BIG CAT MALT LIQUOR	PABST BREWING	
BLACK HORSE ALE	BLACK HORSE BREWING	
BLACK LABEL (NEW)	CARLING BREWING	
BLACK LABEL EXPORT MALT LIQUOR	CARLING BREWING	(2 FOR 1)
BURGEMEISTER	PETER HAND BREWING	
BUSCH BAVARIAN (14 OZ)	ANHEUSER-BUSCH BREW	
CHAMPAGNE VELVET	JOS S PICKETT & SONS	
CINCI	O'KEEFE BREWING	(2 FOR 1)
COLUMBIA	CARLING BREWING	(2 FOR 1)
COOK'S	G. HEILEMAN BREWING	
CORONA CERVEZA (10 OZ)	CERVECERIA CUAUHTEMO	(2 FOR 1)
COUNTRY CLUB	PEARL BREWING	
COUNTRY CLUB MALT LIQUOR	PEARL BREWING	
DIXIE	DIXIE BREWING	
DRUMMOND BROS	FALLS CITY BREWING	
DUBUQUE STAR	JOS S PICKETT & SONS	
DUQUESNE BAVARIAN	DUQUESNE BREWING	
EDELWEISS	G. HEILEMAN BREWING	
EINBOCK BOCK	WALTER BREWING	(2 FOR 1)
ESQUIRE	JONES BREWING	
ESSLINGER	RUPPERT BREWING	
FALSTAFF (14 OZ)	FALSTAFF BREWING	
FINAST (NO BOTTOM)	EASTERN BREWING	(2 FOR 1)
FISCHER'S	FISCHER BREWING	
FISCHER'S ALE	FISCHER BREWING	
FOSTER "E" LAGER (13 OZ)	CARLTON & UNITED LTD	(2 FOR 1)

COMPUTERS are used to keep track of collections and, in some cases, to prepare trading lists. Here is Roy Fuerherm's computer printout showing the cans he has for trade.

star by the cans you've got in the BCCA Composite List. Others use the BCCA's *Guide to United States Beer Cans* in the same way. (Unfortunately, it's only the first of many planned volumes and, as such, is incomplete.)*

Richard Roser, an Ohio teacher, has developed an effective system of record keeping he recommends highly. "I make a 3 x 5 card for each can in my collection," he says. "I have a regular format which includes the brand name, can size, brewery name, description of can, date I obtained it and from whom—and whether or not the can is obsolete. I use white cards for U.S. pull-tabs and flattops, mustard color for U.S. cones, pink for all foreigns, and blue for all U.S. 16 ouncers."

To really put the finishing touches on his cards, Roser has gotten a second BCCA Guide and is gluing the picture of each can on the back of its respective card.

*(St. Louis: Beer Can Collectors of America, 1975).

Several collectors who work with computers by day have these very same computers keeping track of their collections by night. When Ron Moermond, a computer department head, is in doubt about whether or not he has a can, he just pulls out his 2 inch thick printout. It lists all 4500 plus cans in his collection for easy reference.

The ultimate, of course, is to have actual photographs of your entire collection to refer to. The advantage of this method is that you not only can check small details of any can, you can also check its condition. The disadvantage—besides the expense—is that it requires constant updating.

Photographing cans, whether for record keeping purposes or to show off your traders, is tricky business. Many cans—especially the ones that have large areas of metallic paint—reflect light like crazy. For good can photography, then, you cannot use a flash or any conventional lighting setup. The best solution is to wait for an overcast day, then set up your cans outdoors. This even light coming from all directions shows off all cans to their best advantage.

Besides keeping track of what cans they've got, a number of collectors like to note—right on the can—who they got it from. Dick Roser prints the date of acquisition and from whom it was acquired on the bottom of each of his cans with a Sharpie marks-on-anything pen. (Most other felt-tip pens will readily smear.) "Should I ever wish to remove this lettering," says Roser, "it is easily done with lacquer thinner on a Q-tip."

Pressure-sensitive labels are also good for this purpose and quite widely used.

Baltimore collector Charles Miller accomplishes the same thing with a simple piece of tape. "To me, this is very important," he says. "The personalities in my collection are as important as the cans."

Amen, brother.

19

Watch out for fakes

"I think I've been ripped off."

Clint Leonhardt, Beer Can Collector
Louisville, Kentucky

Any day you can trade for a conetop is a good day, indeed.

That's what one collector used to think, anyway. Until he took his Gettelman cone (mint, yet!) and decided to remove the cap. He knew the cap wasn't original, so he wanted to do away with it.

Boy, was he in for a surprise!

Not only did the cap come off—so did the entire conetop. What he had traded for became quickly obvious: A fairly recent Gettelman flattop with a phoney conetop.

Fake conetops are appearing on the scene in steadily-increasing numbers. Only a smattering have shown up at bona fide trading sessions, but they're frighteningly common at flea markets. Maybe it's because it's so easy to attach the top from a Heet or STP can to a common flattop can. Or maybe it's because beer can collectors are super-gullible. Whichever it is, watch out!

SURPRISE! The mint conetop isn't really a conetop at all. It's a mint flattop plus the cone off an automobile additive can.

Some flea market dealers are semi-honest about the whole thing. They'll label their fake cones as novelty items and even go so far as to tell the novice collector something like this: "Sure, they're fakes, sonny. But here's a chance to have some conetops in your collection for only five bucks each!" If this is questionable, what other dealers do is even more so. These crooks tell collectors that the cans they have for sale—often currents-plus-conetops—are experimental, one-of-a-kind jobbies.

For anyone who's building a collection of phoney beer cans, these may be real finds. But for collectors of *real* beer cans, they're very much to be avoided.

The same goes for reprints of scarce old cans that are beginning to trickle out into circulation. If you're into collecting reproductions, that's great. But heaven help anyone who passes one off as being the real thing. Examine any suspected reprint carefully to see if it says REPRODUCTION on it. An honest manufacturer of reprints would so identify them. If they didn't, they'd be counterfeiters.

Reprinting beer cans requires a substantial investment, of course. To do it requires access to sophisticated lithography equipment, not something you're likely to find around the house.

Another type of beer can fakery, however, can very easily be accomplished in a home workshop. (Yes! You, too, can make beer cans at home in your spare time!) All you need is a supply of beer cans (that'll be easy because any brand will do), some masking tape, a can of spray paint, a little glue, and a few old beer bottle labels (these, too, are not difficult to come by.)

After masking off the rims, you paint the cans one solid color. After they dry, you merely glue on your old beer labels, one or two to a can. Voila! You've got some beer cans nobody's ever seen the likes of before! They ought to trade like hotcakes!

And you ought to be ashamed of yourself.

A COLLECTOR GOT RIPPED OFF when he traded for this one. Can is actually a sodapop can that has been painted red, then had a beer bottle label pasted on it.

FROM THE PAINTBRUSH of Steve Peratt came these fictitious brands of beer. It's one way to have cans in your collection nobody else has!

Making phoney beer cans that don't pretend to be anything other than phonies, on the other hand, can be lots of fun.

Omaha's Steve Peratt has taken paint brush in hand and turned out a number of one-of-a-kinders just for his own amustment. Ever heard of American Eagle or Bicentennial Malt Liquor? How about Ptui? Of course not. Peratt has the only ones there are. Or ever will be.

Mass-producing-just-for-fun cans can easily be done, too. Merely have your original design printed up on $8\frac{1}{4}$" x $4\frac{9}{16}$" sheets of paper, then wrap them around old beer or soft drink cans. (You'll need to have your labels made $8\frac{1}{16}$" x 5" if you're going to use the taller, narrower extruded cans.)

PAPER LABEL CANS made just for the fun of it.

The paper label gambit has been used many times, but the two best-known examples are Tree Frog and Srooc.

Tree Frog, "The Sleezy People's Beer," is said to have been a promotional idea of a midwest radio station in the early 1970's. But you'd never know it from the label. All the information it provides is the legend of Louella, the Queen of Sleeze, very likely a take-off on Olde Frothingslosh—which, as was mentioned, is a take-off on the Miss Rheingold can. Whew!

Srooc, which is Coors spelled backwards, was sold during the BCCA Canvention in Denver by Duffy's Shamrock Tavern as a tip of the hat to Canventioneers. The three-color label came wrapped around cans of conventional Coors and was a dead-ringer for the real thing except for the strange brand name it bore.

Nobody would ever mistake Tree Frog or Srooc for real beer cans (or trade anything for them), but collectors do have to be on the lookout for cans that represent themselves to be something they really aren't.*

After all, there *are* a few unscrupulous characters around who will go to almost any length to get their hands on your older and rarer traders.

May all their beer be flat. And all their beer cans be rusty.

*On occasion, brewers have used paper labels for emergency or test purposes. These cans are, of course, are legitimate beer cans. Examples: Primo, Burgemeister, Schlitz Malt Liquor.

20

The canventions

> "The collectors came from Colorado, Canada and Connecticut. They came in campers, Cadilacs and Camaros. But they had one thing in common: Every single car was packed tight with empty beer cans."
>
> *New York Times*

Want to add a quick 100 cans to your collection?

No problem. Next September, just go where the really concentrated trading takes place. To the Canvention. Rub elbows and trade with collectors of all ages, who have all sizes of collections.

Of course, you may not actually pick up a hundred new cans for your collection. You may well wind up with many more than that. Frank Visconti sure did. He came to a Canvention with a mere 30 traders. And took home 243 cans for his shelves.

If collecting empty beer cans is dumb, traveling halfway across the country to trade them is even dumber. But it happens every September.

ONE DAY, 13-year-old Shawn Poteet decided to find enough beer cans to make one of those beer can hats. Before the day was over, he had gone through enough of Iowa's roadside barrels to have 21 different brands—and he was hooked. He forgot all about the hat and dedicated his life to beer can collecting. Next thing he knew, his family had, too. And pretty soon Shawn found himself at a Canvention, trading cans with bigtime Wyoming collector, Jeff Berg.

TRADING. That's what the Canvention is all about. Here, Dr. J.D. Kerr (a veterinarian from Decatur, Illinois) trades for some cans with Curt and Lance Paulsen of St. Louis. (Bulldogs and Big Cats, we'll bet!)

MISS BEER CAN. That's why Chris Jennings was at Canvention V in Des Moines. And everybody wanted to have his picture taken with her. Like Max Robb, a beer can fancier from Central City, Colorado.

THE FIRST FIVE commemorative cans produced for the BCCA Canventions. The three on the left were silk-screened. The other two were lithographed.

It began in September of 1971 in St. Louis. The BCCA, then a mere 274 members strong, decided to see what would happen if they invited everybody to a big beer can emptying and trading session in the courtyard of the Holiday Inn.

What happened is that over half of the members showed up from 17 states plus Canada. More than twice as many collectors showed up for Canvention II the following year. Hamm's, Miller and Schmidt donated a total of 157 cases of you-know-what for the occasion, and by the time Canvention III rolled around, 563 collectors arrived at Cincinnati's Netherland-Hilton.

September of 1974 heralded a National Canvention in Denver. Collectors showed from 34 states. Plus Canada. *Plus* the Canal Zone. There was even a daily newspaper put out by the Mile Hi Chapter.

And there was a crisis, too. After completing a trade on another floor of the hotel, John Zembo returned to his room and noticed that there was something missing. A case of 35 cans of missing obsoletes! He called housekeeping, room service, maid service—everybody. Finally, the security chief heroically found them. Neatly stacked in garbage containers. Whew!

Member #1646 Beer Can Collectors of America
James Mitchell
Collector of Beer,
Ale, & Malt Liquor Cans.
1307 Devonshire Dr. 815
Joliet, Illinois 60435 729-0036
"OPEN ALL CANS FROM THE BOTTOM, IF POSSIBLE."

Martin Landey, no. 78
Collector of fine beer cans and beer bottles and the whole world of breweriana
80 Fairview Ave.
Belmont, MA. 02178
(617) 489-3527

Cowboy Chapter - B.C.C.A.
JEFF BERG No. 799
(307) 324-9540
1627 WEST MAPLE, APT. N
RAWLINS, WYOMING 82301

WORLD'S LARGEST BEER CAN COLLECTION
— — More than 8000 different cans — —
WANTED: old, new, foreign, U.S., mint, dump, ANY BEER CANS
JOHN F. AHRENS
192 RAMBLEWOOD PARKWAY
MT. LAUREL
MOORESTOWN, NEW JERSEY 08057
609-235-2496
BCCA No. 9, DIR., 1971 - 1974
CHAIRMAN, PHILA. '76 CANVENTION
SEPTEMBER 9 - 12, 1976
ECBA No. 14

334 HIGHLAND AVENUE
WORTHINGTON, OHIO 43085
614-885-5041
BCCA 2404
EMPTY BEER CANS - WORLDWIDE
THE WORTHINGTON BEER CAN GAL
JO-ANN McCLURE
AND HUSBAND BOB
MEMBER
BEER CAN COLLECTORS OF AMERICA

"Your Trash May Be My Treasure"

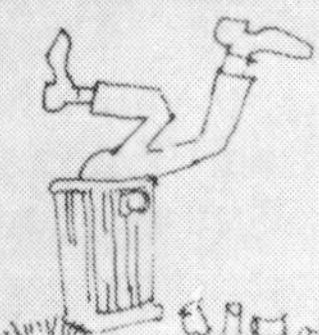

GARY L. FRONK
BEER CAN COLLECTOR
Member
BCCA - WWBCC
3013 Wolcott
Des Moines, Iowa 50321
(Home) 515 285-0403
(Office) 515 225-2000

SAVE THAT CAN — LET ME THROW IT AWAY!
CHARLIE MILLER
THE ORIGINAL
BEER CAN SCAVENGER

2722 SUPERIOR AVENUE
BALTIMORE, MD. 21234
(301) 665-2464
SUPPORT YOUR LOCAL BREWERY
Member: B.C.C.A. (30), E.C.B.A., F.& A.M., E.P.C., B.P.O.E, F.O.E.

"The Origional Cincinnati Beer Can Collector"

Buy Local Beer — Keep'em Brew'in
JOHN P. PAUL
BCCA #42
513-921-5468
513-922-6756
809 Depot St.
Cincinnati, O. 45204
OVER

THE ORIGINAL LITTLE ROUND MAN FROM ST. LOUIS SAYS-

"THE LIFE OF A BEER CAN COLLECTOR IS NOT AN EASY ONE. IT SEEMS SOMEONE IS ALWAYS TRYING TO RECYCLE YOUR CAN."

HAL LEEKER B CCA 843

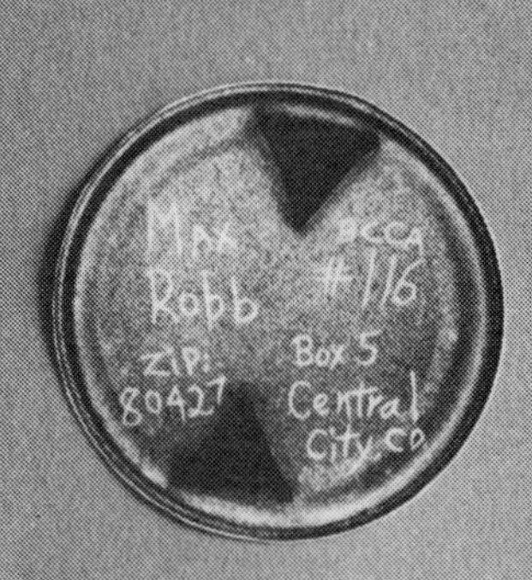

Used Cars – Hogs – Whiskey – Manure – Nails – Jade
Flyswatters – Racing Forms – Bongos – Used Golf Balls

Larry Wright
BCCA #2

380 Pebble Acres Dr. St. Louis, Mo. 63141 (314) 434-3755
Beer Can Collector, Cotton Picker and Entrepreneur

Flats Fixed	Possum Hunts Organized	Uprisings Quelled
Bars Emptied	Squirrels Skinned	Horses Broken
Fish Cleaned	Exhorcisms Performed	Tigers Tamed
Pedicures	Septic Tanks Cleaned	Excuses Verified
Feuds Started	Acupuncture Reasonable	Orgies Organized
Trash Hauled	Circumcisions Cheap	Rooms Rented

CHARTER MEMBER - BEER CAN COLLECTORS OF AMERICA

CANVENTION TIME brings hundreds of collectors together from Kan-sas, Can-ada, Kan-tucky—and everyplace in between.

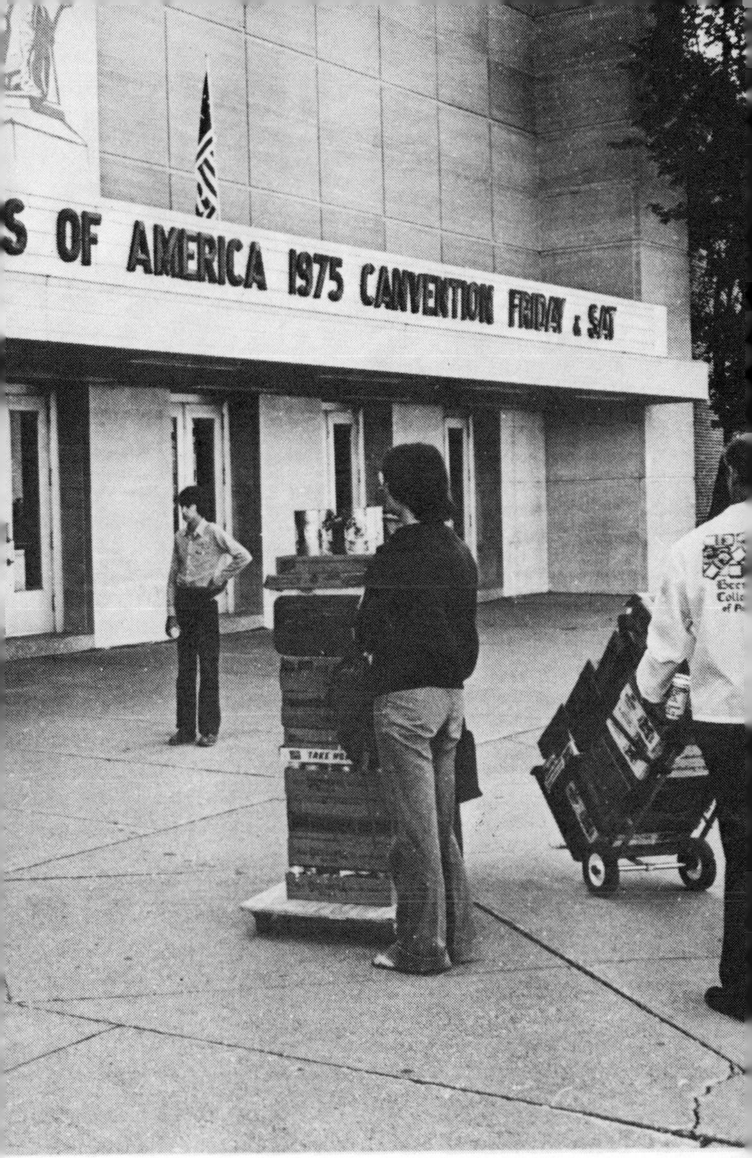
S OF AMERICA 1975 CANVENTION

THE TRADING SESSION at a BCCA Canvention. With all these cans available for trade, any collector—beginner or advanced—is sure to pick up some he didn't have before.

Steelers

At Canvention V in Des Moines, it took several hotels to hold all the can collectors. This Hawkeye Chapter-sponsored get-together got together a total of 1,023 members (plus 770 guests). An estimated half-a-million traders were lugged, hauled, towed and flown into Des Moines for the occasion. (Collector Bob Leslie discovered that exactly 104½ cans would fit in his station wagon.) Trading sessions had to be held in Veterans Memorial Auditorium—and even that giant facility was taxed by the immense number of collectors (and cans).

The next Canventions would be at Philadelphia ('76), Kansas City ('77), and Milwaukee ('78), but it was at Des Moines—Canvention V—that the whole concept really came of age.

From the courtyard of a motel to a huge exhibition hall—in just five years. If that doesn't say something about the "beer can collecting craze" (as *The Wall Street Journal* has called it), nothing will.

O.K., you've decided to go. Besides registering and arranging for rooms and all that, what do you need to know?

1. When to come: The early arrivals *do* get some mighty fine trades. Best to plan to arrive Tuesday or Wednesday.

2. What to bring: Don't bother with fancy duds. Nobody's interested in trading clothes with you. Just bring a couple hundred of your best traders—and a churchkey to bottom-open all the beers you'll be offered by a couple hundred of your better collectors. (There is no need to bring 90 cases of traders like John Udvare brought to Canvention II. Unless you're John Udvare and have a reputation to uphold.) You won't want to bring any cans from your collection, either, unless you plan to enter one of the display competitions. (And they're always limited to 100 cans.)

CANVENTION GARB often is very beery. Jack Isacson and a friend model the latest in collectors' apparel.

3. How to maximize your trading power: Spend half your time in your room waiting for good trades to drop in. And spend the other half visiting other rooms with an armload of traders.

INGENIOUS METHODS of toting your cans-for-trade around at a Canvention are important if you're going to add another 100 cans to your collection.

(This means you'll need a good way to carry a bunch of cans around with you. Let your ingenuity be your guide.)

4. What to do when the metal detector at the airport goes crazy over your luggage: Relax. Airport security people are starting to get used to people who carry three cases of empty King Snedley's in their suitcases.

There are about fifty collectors who have been to every Canvention. And a competition has developed among them to see who will be the last who can truthfully say he's attended Every Canvention Ever Held.

Because of his callow youth—and his callous attitude—Bill Christensen is likely to be the eventual winner. "I *will* be the last one to have attended all BCCA Canventions." he states. "If there ever comes a time that I cannot attend, I will somehow arrange to have it postponed. Perhaps I'll burn down the hotel the weekend before the Canvention is scheduled. This is only right and proper."

Other than that, he's an O.K. guy.

21

The breweries

"It's strange, but we Americans—one of the most adventurous peoples in the world—are sometimes downright unimaginative when it comes to our beer."

Chuck Nekvasil, Beer Can Collector
Cleveland, Ohio

America's top ten brewers sell 80% of all the beer America drinks. (The Big Four alone sell 56%!)

That leaves only 20% to be divided up among the other 47 or so brewing companies. And those numbers—both the 20% and the 47%—are getting smaller all the time.

The plight of the small breweries is a very real one. It's brought on, to some degree, by the American tendency to associate bigness with goodness. (But what about the American tendency to support the underdog?)

Of course, technology had much to do with it, too. Back in the old days, a brewer didn't have to concern himself with pasteurization, refrigeration or transportation. All beer was made locally and nobody even considered transporting it hundreds of miles. How could they? It would go bad. With today's mass production, mass marketing and mass distribution, it isn't surprising that the little guys are getting forced out.

Besides, they are often saddled with old, inefficient plants and their costs are continually rising while the big outfits are automating and reducing costs—or at least holding the line better.

"Today," says one industry analyst, "there is no way for a local brewer to survive unless he gets a grant from the Ford Foundation."

Fat chance. There were 750 brewers in the U.S. when the nation went wet again in 1933. By 1950, there were only 407. And by 1971, there were only 76 left. At last count, there were less than 50.

In recent years, the fires have gone out under brew kettles in Cleveland, Oklahoma City, San Francisco, New Orleans, Omaha, Pueblo, New York City and Huntington, West Virginia. Wisconsin (which once had 226 breweries) has had two more closings of late—in Potosi and Rice Lake—and is now down to eight brewers. Chicago (which had 11 breweries as recently at 1958) is now down to one.

Associated Brewing thought they had the answer to the brewing biggies when they brought together a lot of regional brands under one corporate umbrella including Piel's, Pfeiffer, and Drewrys. Matter of fact, as recently as 1966, Associated ranked as the 7th largest brewer in the country.

But these smaller brands just couldn't attract new customers. Nor could they keep their old ones from drifting away to the Buds and Pabsts of the world.

In 1970, Associated closed the first of its breweries. In 1972, they sold three other breweries to Heileman. In 1973, they sold their last two breweries—the Piels plants in Massachusetts and New York. And that was the end of Associated. From Number 7 to zero in less than seven years.

Support your local brewery! That's the cry of the beer can collector. But there are fewer and fewer to support all the time.

Here, then, is a list of the brewing companies that remain in the beer making (and canning) business and the brands they've canned in the last year or so that you shouldn't have too much trouble adding to your collection. (Not listed: The irregularly-produced private label brands.)

If this list becomes smaller—and, unfortunately, it probably will—at least you'll know you have some obsolete cans to remember the defunct brewers by. If, on the other hand, the smaller brewers do manage to hang on, you can be assured of a continuing supply of new brands and designs for your collection.

ANHEUSER-BUSCH:	(multi-location)	Budweiser, Busch Bavarian, Michelob.
BLITZ-WEINHARD:	(Oregon)	Blitz-Weinhard, Cascade, Alta, Acme, Buffalo.
CARLING:	(multi-location)	Carling Black Label, Tuborg, Heidelberg, Columbia, Red Cap Ale, Columbia, National Premium, National Bohemian, Colt 45, Van Lauter, Altes, A-1.
CHAMPALE:	(N.J. and Virginia)	Champale, Black Horse Ale, Regent, Old Dutch Beer.
COLD SPRING:	(Minnesota)	Cold Spring, Fox DeLuxe, North Star, Kegle Brau, Karlsbrau, Western, Northern, Gluek, Gemeinde Brau.
COORS:	(Colorado)	Coors.
DIXIE:	(Louisiana)	Dixie, Dixie Light.

DUNCAN:	(Florida)	Dunk's, Fischer's, Fischer's Ale, Regal.
EASTERN:	(New Jersey)	Milwaukee Premium, Canadian Ace, Old Bohemian, Fox Head "400," Old German.
FALLS CITY:	(Kentucky)	Falls City, Drummond Bros.
FALSTAFF:	(multi-location)	Falstaff, Narragansett, Krueger, Krueger Pilsner, Krueger Ale, Narragansett Ale, Hanley, Ballantine, Ballantine Ale.
GENERAL:	(California and Washington)	Falstaff, Lucky Lager, Lucky Red Carpet, Lucky Dark Continental, Fisher, Spring, Ballantine, Heritage House, Maier, Keg, Golden Crown, Lucky Draft, Lucky Bock, Brew 102, Regal Select, Padre, Hof-Brau, Tivoli, Springfield, Reiderbach, King Snedley's.
GENESEE:	(New York)	Genesee, Genesee Cream Ale, Fyfe & Drum.
PETER HAND:	(Illinois)	Old Chicago, Old Chicago Dark, Braumeister, Old Crown, Old German, Old Crown Ale, Alps Brau, Van Merritt, Burgemeister, Oertels, Peter Hand Extra Light, Oertel's '92.

HEILEMAN:	(multi-location)	Blatz, Sterling, Special Export, Wiedemann, Pfeiffer, Mickey's Malt Liquor, Drewrys, Old Style, Schmidt, Kingsbury, Grain Belt, Storz, Havenstein, White Label.
HORLACHER:	(Pennsylvania)	Horlacher.
HUBER:	(Wisconsin)	Huber Premium, Hi-Brau, Wisconsin Club, Our, Wisconsin Gold Label, Regal Brau, Bavarian Club, Rhinelander, Bohemian Club, Holiday, Alpine.
HUDEPOHL:	(Ohio)	Hudepohl, Hofbrau, Burger, Tap.
HULL:	(Connecticut)	Hull's Export.
JONES:	(Pennsylvania)	Stoney's, Esquire.
KOCH:	(New York)	Koch's, Koch's Golden Anniversary, Iroquois Craft, Bavarian Select.
LATROBE:	(Pennsylvania)	Rolling Rock.
LEINERKUGEL:	(Wisconsin)	Leinerkugel's, Chippewa Pride, Bosch, Gilt Edge.
THE LION:	(Pennsylvania)	Gibbons, Bartels, Stegmaier Bock.
LONE STAR:	(Texas)	Lone Star, Buckhorn.
MILLER:	(multi-location)	Miller, Milwaukee's Best, Meister Brau, Lite.
MT. CARBON:	(Pennsylvania)	Bavarian Type.

OLYMPIA:	(Minnesota and Washington)	Olympia, Hamm's, Buckhorn.
ORTLIEB:	(Pennsylvania)	Ortlieb, Kaier, Ivy League, Old English 800.
PABST:	(multi-location)	Pabst, Andeker, Old Tankard Ale, Pabst Bock, Burgermeister, Red, White & Blue.
PEARL:	(Missouri and Texas)	Pearl, Pearl Light, Goetz, Country Tavern, Country Club, Country Club Malt Liquor, Texas Pride, Jax, Brown Derby.
PICKETT:	(Iowa)	Pickett's Dubuque Star, Fox Head, Edelweiss, Champagne Velvet.
PITTSBURGH:	(Pennsylvania)	Iron City, Tech, Mustang Malt Liquor, Hop 'n Gator, DuBois, DuBois Export, Gambrinus, Gambrinus Gold, Augustiner, Mark V, Old German, Old Dutch, American, Old Export, Olde Frothingslosh, Brown Derby, Seven Springs.
RAINIER:	(Washington)	Rainier, Rainier Ale, Rheinlander.
READING:	(Pennsylvania)	Reading, Bergheim.
RHEINGOLD:	(multi-location)	Rheingold, Knickerbocker, Jacob Ruppert, Esslinger, Gablinger's.

SCHAEFER:	(multi-location)	Schaefer, Gunther, Piels.
SCHELL:	(Minnesota)	Fitger, Stein-Haus.
SCHLITZ:	(multi-location)	Schlitz, Old Milwaukee, Schlitz Malt Liquor, Primo, Schlitz Light.
SCHMIDT:	(Ohio and Pennsylvania)	Schmidt's, Silver Top, Duke, POC, Valley Forge, Prior, Kodiak Ale, Rams Head Ale.
SCHOENLING:	(Ohio)	Schoenling.
SPOETZL:	(Texas)	Shiner.
STEVENS POINT:	(Wisconsin)	Point Special.
STROH:	(Michigan)	Stroh's, Goebel.
WALTER:	(Wisconsin)	Walter's, Bub's, Master Brew, Old Timer's, Bruenigs.
WEST END:	(New York)	Utica Club, Utica Club Cream Ale, Matt's, Maximus Super.
YUENGLING:	(Pennsylvania)	Yuengling's.

A number of brewing companies are owned by other brewing companies. (Stag by Carling, Burger by Hudepohl, Wiedemann and Sterling by Heileman, for example.) Whenever this is the case, we have listed only the parent brewing company.

No breweries are listed here which produce only bottled beer. That's why Anchor Steam is not mentioned anywhere in this book.

IMPORTANT: Do not bug the breweries listed above for cans—or anything else. Used to be, when there were only a handful of collectors, they were happy to answer collectors' letters. (Some would even send free cans!) *But no more.* By now, they've all suffered mightily from a barrage of collector mail.

Besides, when you bother the little breweries, you're just pushing them closer to the edge. They just plain can't afford to answer requests from hobbyists. So don't bother them.

Just drink their beer.

"If there are any old beer can collectors out there about ready to die, don't forget me in your will. I'll give your beer cans the love and tender care they deserve."

Van Wagner, age 16, Beer Can Collector
Galion, Ohio